Introduction to Surface Physics

Introduction to Surface Physics

M. PRUTTON

Department of Physics,
University of York

Clarendon Press · Oxford · 1994

Oxford University Press, Walton Street, Oxford OX2 6DP

Oxford   New York   Toronto
Delhi   Bombay   Calcutta   Madras   Karachi
Kuala Lumpur   Singapore   Hong Kong   Tokyo
Nairobi   Dar es Salaam   Cape Town
Melbourne   Auckland   Madrid
and associated companies in
Berlin   Ibadan

Oxford is a trade mark of Oxford University Press

Published in the United States
by Oxford University Press Inc., New York

A catalogue record for this book is available from the British Library

Library of Congress Cataloging in Publication Data
Prutton, M.
Introduction to surface physics/M. Prutton.
1. Surfaces (Physics)   2. Surface chemistry.   I. Title.
QC173.4.S94P78   1994   530.4'17—dc20   93–31100

ISBN 0–19–853475–2 (hbk)
ISBN 0–19–853476–0 (pbk)

Typeset by
Cotswold Typesetting Ltd, Gloucester
Printed in Great Britain
by Bookcraft (Bath) Ltd
Midsomer Norton, Avon

Preface

This book is a new version of an introductory text on surface physics which is intended for final year science undergraduates and new graduate students. Eighteen years have passed since I wrote the first edition and it is about 30 years since research in this field began to grow explosively. Not surprisingly, the literature of the subject is gargantuan in size and broad in content. Therefore the material chosen for inclusion here is a very small subset of the variety of topics to be found in textbooks, research monographs, and the scientific literature. This subset is used in a lecture course to final-year physics undergraduates and is intended not to be so comprehensive that the student can read it and go away and start research, but rather to act as a broad, and I hope stimulating, introduction to a large subject.

Thus, an attempt has been made to describe why surface studies are important scientifically and technologically, what techniques are available and how they compare with each other and with bulk methods, and what types of problems have been and are being tackled. The emphasis upon technique so evident in the first two editions has been reduced by adding material about some of the results that have been obtained and about some of the theoretical advances that have been key in helping provide understanding of the wide variety of phenomena that occur at surfaces. Chapters 2 and 3 are concerned mainly with techniques for the determination of what kinds of atoms are present on a surface and in what quantities, and how these atoms are arranged with respect to one another and the underlying material. Chapters 4, 5, and 6 then deal largely with selected case studies and particular systems chosen so as to illustrate the application of the techniques to problems concerning, respectively, the electronic, vibrational, and adsorption properties of surfaces. The particular case of the (111) surface of elemental silicon pervades all the material because it has been the subject of a huge worldwide scientific effort, owing to the interesting way in which the surface atoms arrange themselves, the technological importance of this surface in large-scale integrated circuits, and the impressive progress that has been made in understanding how and why the surface and the inside of the solid are so different.

In order to keep this book down to an affordable size for students a very large number of uncomfortable decisions have been made about just what to leave out. This must mean that many readers will be irritated at not finding some topic close to their interests, and I am sorry about that. Nevertheless, I have tried to cover the areas of the subject which set a foundation upon which towers can be built by reading and thinking. When I have curtailed or missed out an area I have tried to give a reference which will lead the reader into it, starting from the level of this book. Indeed, nearly all the references are intentionally made to review articles in learned journals like *Reviews of Modern Physics, Reports on Progress in Physics, Surface Science Reports*, or to other books. This means that I have deliberately not included a bibliography which always reflects fairly on the first researchers to have a particular idea or make a new measurement. These references are in the books and review articles cited. Again, my apologies to all the authors of original papers who are not cited—I hope that you sympathize with the aim of keeping the text to moderate size and cost!

The books on surface physics which are now available usually start at a level assuming some knowledge of solid state physics, and the texts by Kittel (1986) and Rosenberg (1975) are referred to throughout this book and are important aids in this respect. Some of the other texts on surface physics or chemistry which the reader may find useful are Somorjai (1972), Blakely (1973), Woodruff and Delchar (1988), Zangwill (1988), and Briggs and Seah (1990, 1992). Full references to these works are given near the end of this book.

I am indebted very heavily to the kindness and tolerance of my colleagues at York University, who have helped in many ways with the material presented here. In particular I would mention Professor J. A. D. Matthew, Dr A. Chambers, Dr T. E. Gallon, and Dr S. P. Tear in the Department of Physics and Dr M. M. El Gomati in the Department of Electronics, who have helped read and check manuscripts in a most cooperative way as well as teaching me how to study many aspects of surface physics during our long collaboration. I have been privileged to have worked with many first-class graduate students, post-doctoral researchers, and visiting academics during the last 15 years and have learned a great deal from them which has become subsumed into aspects of this text. Particular thanks go to Drs I. R. Barkshire, P. J. Bassett, V. E. de Carvalho, P. G. Cowell, J. C. Greenwood, P. G. Kenny, D. C. Peacock, C. G. H. Walker, and M. R. Welton-Cook, who have worked in my research group. The work on Auger microscopy mentioned in Chapter 2 was strongly influenced by Dr R. Browning of Stanford University and Dr H. Poppa of IBM, Almaden, as well as Dr R. H. Roberts of the University of Newcastle, New South Wales.

Acknowledgements to the many friends and colleagues who provided material for the figures are made in the relevant captions. I am very grateful to Alan Gebbie, who made excellent versions of these figures as well as drawing many new illustrations.

York
January 1993   M.P.

Contents

1
Introduction

All the properties of a piece of bulk material are determined by the number and types of atoms it contains and by their arrangement in space with respect to each other. Some properties can be related in a straightforward manner, both theoretically and experimentally, to the chemical composition and to the crystal structure by using the large body of understanding provided by the band theory of solids (Rosenberg 1975; Kittel 1986). Thus, for instance, the division of crystalline solids into insulators, semiconductors, and conductors, the explanation of the relationship between thermal and electrical properties, and the occurrence of both Hall and magneto-resistance effects can all be satisfactorily explained within the framework of the band theory of solids.

It may be less straightforward to relate other properties to a theoretical model of a solid, and a more empirical approach may have to be adopted. One example of such a property is the ferromagnetism of some metals. This depends upon small differences between large interactions in the solid, and demands a difficult and sophisticated theory for its explanation. Mechanical creep and fatigue failure are examples of phenomena requiring an understanding of faults which occur in crystalline solids and the way in which they move in response to applied forces. Again, the theoretical description of these processes is difficult. Nevertheless these properties are still determined in principle, provided that the composition and structure of the material are sufficiently well defined.

The subject of surface physics is the study of the chemical compositions and atomic arrangements at the surfaces of solids and the theory and observation of their mechanical, electronic, and chemical properties. As in the study of bulk solids, the ultimate objective is the establishment of understanding of the relationships between the properties, the composition, and the structure. There are many reasons for expecting that a solid surface will have different properties from the bulk material, and these provide an incentive for the physicist to enquire and try to understand. Equally importantly, there are many processes of technological significance which depend upon the use of solid surfaces and which may be improved in some way if the role of the surface could be fully understood.

In this book the surface is thought of as the top few atomic layers of a solid. In many older books on surface chemistry or metallurgy the surface is regarded as the top 100 nm or so of the solid. The larger distance was determined more by the techniques that were available at that time than by any more basic physical consideration.

The reasons for the expectation that a surface will have different properties from the bulk of the solid may be understood by considering a surface formed by cutting through the solid parallel to a chosen plane or atoms. If the atoms are not disturbed from their bulk equilibrium positions by this operation then the surface can be said to be a bulk exposed plane (Fig. 1.1(a)). Such a plane shows the minimum disturbance of the solid arising from the formation of the surface.

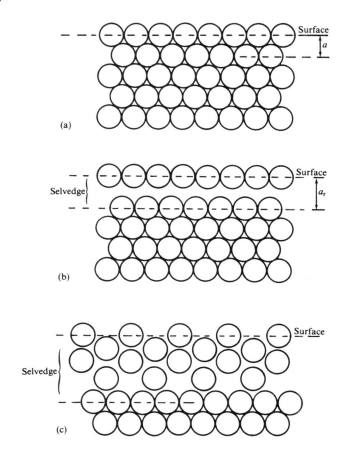

Fig. 1.1 Rearrangement of atomic positions at a solid surface. Haxagonal close-packed atoms. (a) The bulk exposed plane; (b) relaxation of the surface plane outwards; (c) reconstruction (hypothetical) of the outer four atomic planes.

Even so, because many electronic properties of the bulk depend upon the three-dimensional periodicity of the potential inside the solid, the loss of periodicity in one dimension due to the existence of the surface will result in a change in the electronic states near and at the surface, and so surface electronic properties differ from bulk (Chapter 4). Also, the lack of nearest neighbours on one side of the surface atoms may make available chemical bonds which 'dangle' into the space outside the solid and which will be available for chemical reactions (Chapter 6).

It is more likely that the disturbance caused by terminating the solid in a surface, particulary the disturbance due to the absence of the bonding force of nearest neighbours on one side of the surface atoms, will result in new equilibrium positions for the atoms in and near the surface. The simplest change of this kind is the relaxation illustrated in Fig. 1.1(b). Here, the separation of the surface plane from the next plane of atoms happens to be drawn so as to be greater than the corresponding separation in the bulk solid. This deviation from the bulk spacing may continue, in decreasing magnitude, as one probes deeper into the solid. The surface region over which there is a deviation from the bulk lattice spacing is referred to as the selvedge. Relaxation retains the symmetry of the atomic arrangement parallel to the surface but changes the spacings normal to the surface. It may result in changed properties for the surface because, for instance, it could create an electric dipole moment in the selvedge. A more extreme disturbance occurs when the surface atoms rearrange themselves into a structure with symmetry altogether different from the bulk solid. This phenomenon is called reconstruction, an example of which is shown in Fig. 1.1(c). This reconstruction modifies the symmetry near the surface and will affect all the structure-sensitive properties of the surface—the atomic vibrations, and chemical, optical, and electronic behaviour.

Many different kinds of processes which are of great scientific and technological interest occur at surfaces. The variety among these processes is very large, a fact which accounts for the wide spread of disciplines involved in surface physics. A few of them are listed below in order to give some view of the incentives for surface investigations.

1. *Thermionic emission*. By raising the temperature, sufficient kinetic energy can be imparted to electrons at the top of the conduction band in a metal for them to be ejected from the surface into the vacuum. This process is known as thermionic emission and is important in many electronic devices and most particularly for the source of electrons in oscilloscope tubes and electron microscopes. The number of electrons which can be obtained by thermionic emission is a function not only of the material but also of the presence of chemical contaminants (the cleanliness) of the emitting face and of its crystallographic orientation. The attempt to understand the factors controlling electron emission is an important part of surface physics which is described at greater length in Chapter 4.

2. *Crystal growth*. The development of methods for growing large single crystals of a wide variety of solids has been crucial to the extension of solid state physics from the simplest band models and to the ability of electronic engineers to design semiconducting devices. The processes of crystal growth

generally involve the deposition of atoms upon single-crystal surfaces under such conditions that the arriving atoms can diffuse about and build up the three-dimensional periodic array. Thus, the physics of the energetics and kinetics of adatoms upon single-crystal surfaces is fundamental to an understanding of crystal growth. Some of this type of surface physics is described in Chapter 6.

3. *Chemical reactions.* Many chemical reactions involve interactions between different kinds of atoms across a surface or interface. Even the simplest processes, when viewed at an atomic level, are not fully understood in these terms. One particularly important example is the corrosion of metals or a simple extreme case of corrosion–oxidation. The way in which a clean metal surface is converted to a bulk oxide when exposed to an atmosphere of oxygen must be understood before interpretation of corrosion in more complicated atmospheres is unambiguous. Some simple examples of oxidation on low-index metal faces are described in Chapter 6.

4. *Catalysis.* The presence of surfaces of a particular metal during a chemical reaction can sometimes cause marked increases in the speed of the reaction (Bond 1974). This catalytic action is technologically important, but is the subject of a largely empirical literature. It is one of the longer-term aims of surface studies to throw some light upon the way in which complex practical catalytic systems operate, particularly with a view to finding more economic catalysts than metals like platinum.

5. *Colloids.* Micrometre-sized particles of a solid suspended in a liquid—a colloidal suspension—form an interesting and useful chemical system. Many of its special properties arise from the large surface area of the particles and an understanding of its behaviour must rest upon a knowledge of the role of this surface.

6. *Semiconductor interfaces.* Many semiconducting devices depend crucially upon phenomena that occur at a surface or interface. A junction between p-type and n-type material; a junction between a metal oxide and a semiconductor (MOST devices); the junction between a metal contact and a semiconductor—all three involve the formation of a surface and the preparation upon it of another material. The chemistry and structure of the surface and the way in which these change as the second material is added will affect the electronic properties across the interface. Some of these matters are decribed in Chapter 4.

7. *Brittle fracture.* Some metals and alloys have enormous mechanical strengths when under continuous load. However, they can often be broken by the sudden application of a much smaller load; this phenomenon is called brittle fracture and it can be quite an embarrassment! It appears to be due to the migration of impurity atoms to the grain boundaries in a solid, which

became weak regions under impact. The application of surface techniques to the study of this impurity segregation at grain boundaries may help to provide understanding of the problem and may even lead to the discovery of means for inhibiting the diffusion and so preventing brittle fracture in some materials.

Three different factors have played a part in the rise to the current level of interest in surface physics. In the first place, the theory of both the electronic band structure and chemical bonding in simple bulk solids has been sufficiently successful that theoreticians and experimentalists have been encouraged to attempt to extend the theory. This extension is being explored in two directions simultaneously—the properties of more complicated ionic and molecular bulk solids and the properties of defects in solids. Among the defects the most obvious is the occurrence of a two-dimensional surface bounding a three-dimensional periodic structure. Secondly, the technological pressures mentioned above have become more urgent, and, because techniques became available which could throw light upon the relevant problems, interest has grown in trying to use some of the surface physics as it evolves.

The third is technical rather than historical, but it is nevertheless crucial. It is the development, in association with the interest in space research, of techniques for the achievement of ultra-high vacuum (UHV). This level of vacuum is one in which the rate of impingement upon the surface being studied of molecules from the ambient residual atmosphere in the vacuum chamber is negligible in the time required for the observations. The kinetic theory of gases (e.g. Chambers *et al.* 1989) shows that the number of gas molecules hitting unit area of a surface per unit time is given by:

$$Z = bp(M_r T)^{-1/2}, \tag{1.1}$$

where p is the ambient gas pressure above the surface, M_r is the relative molecular mass of the gas, T its thermodynamic temperature and b is a universal constant. It is usual to write the incident flux density Z in the units molecules $m^{-2} s^{-1}$, and consequently the value of b is written in one of the forms:

$$b = 2.63 \times 10^{24} \text{ molecules } m^{-2} s^{-1} K^{1/2} Pa^{-1}$$

$$b = 3.51 \times 10^{26} \text{ molecules } m^{-2} s^{-1} K^{1/2} Torr^{-1}$$

$$b = 2.63 \times 10^{26} \text{ molecules } m^{-2} s^{-1} K^{1/2} mbar^{-1}. \tag{1.2}$$

All three pressure units (Pa, mbar, and Torr) are currently in use, and hence b is given here for each of these units.

In a conventional vacuum sytem using diffusion pumps and elastomer gaskets the pressure is normally about 10^{-6} Torr and eqn (1.1) shows that this corresponds to approximately 3.0×10^{14} molecules of nitrogen arriving each second on each square centimetre of a surface at room temperature. Since an atomic monolayer corresponds to about 10^{15} atoms cm^{-2} (interatomic distances being about 0.3 nm) such conditions result in nitrogen arrival rates of a monolayer every 3 s, assuming that every molecule sticks to the surface. Since many experiments take longer than a few seconds this represents an unacceptable level of contamination of a surface. Ultra-high vacuum is now generally regarded as the region below 10^{-9} Torr. Equation (1.1) shows that, at room temperature, 10^{-10} Torr corresponds to nitrogen arrival rates of 3.8×10^{10} molecules s^{-1}, or about 1 monolayer in about 8 h. At this pressure the mean free path between collisions of molecules of the ambient atmosphere would be about 50 000 km!

The techniques required to achieve ultra-high vacua are reviewed in many textbooks (e.g. Redhead 1968; Chambers *et al.* 1989). A diagram of the type of UHV system found in many surface physics laboratories is shown in Fig. 1.2, and Fig. 1.3 is a general view of a multiple-technique system in the author's laboratory. The important features of such systems are as follows:

1. The vacuum chamber and its associated pipework are normally fabricated of argon-arc welded or vacuum-braised stainless steel. This material corrodes very slowly and has low rates of outgassing of absorbed gas.

2. The vacuum joints are made with metal instead of elastomer gaskets. Gold o-rings or flat copper rings are normally used here. The use of metal gaskets avoids release of organic contaminants, reduces leakage of water vapour from the atmosphere into the system, and allows the baking described in (3).

3. The whole chamber assembly is desigend so that it can be heated to about 470 K while the vacuum pumps operate. This 'baking' of the system results in accelerated desorption of water vapour (and other gases) from all internal surfaces. When the system is cooled back to room temperature the ultimate pressure attainable is thus substantially reduced.

4. The pumps provided to evacuate the chamber are often ion pumps, titanium sublimation pumps, and, for initial pumping from atmospheric pressure, sorbtion pumps. These three techniques avoid the use of any organic materials. Very well-trapped diffusion pumps filled with a low vapour pressure fluid can be used where special problems arise—an important case being to pump away large throughputs of noble gases, which are only very slowly removed by ion or titanium sublimation pumps. Polyphenyl ether is an example of a suitable fluid for a UHV-compatible diffusion pump and which can be backed by a rotary rather than a sorbtion pump.

5. The choice of materials used inside the vacuum chamber is made carefully to avoid high vapour pressures. Stainless steel, molybdenum and tantalum are in common use for fabricating parts, oxygen-free high-conductivity copper is often used as a conductor, glass and high-density

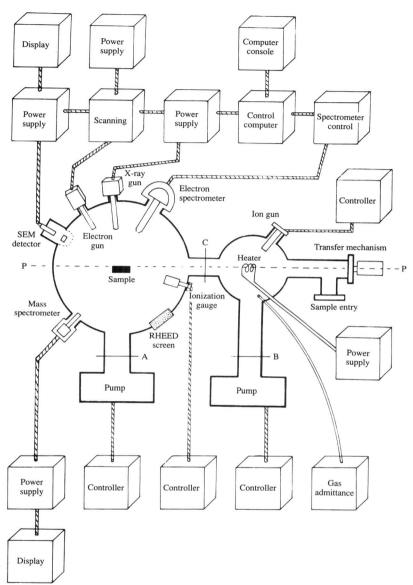

Fig. 1.2 Arrangement of vacuum components in a typical UHV system for surface studies.

ceramics such as alumina (Al_2O_3) are used as insulators, and silver as a material to line bearings because it is such a soft metal.

Properly designed UHV systems reach the low 10^{-10} Torr range on a

Fig. 1.3 A multi-technique UHV system in use for surface studies and equipped for RHEED, mass spectroscopy, Auger electron spectroscopy and microscopy, backscattered electron detection, ion bombardment, and characteristic X-ray detection. The apparatus is surrounded by a cube of coil pairs designed to cancel out the Earth's magnetic field in the region of the sample and the detectors around it.

routine basis, and better pressures are possible with great care. Because of the time required to achieve UHV (10–24 hours) experiments are often designed to allow several *in situ* operations on the sample surface and several *in situ* types of measurement (Fig. 1.3). The sample itself is normally mounted on some kind of manipulator, which may allow various combinations of translation and rotation for alignment and positioning, electrical isolation, heating, cooling, and *in situ* cleavage as a means of preparing a clean single-crystal surface. Further, it is common practice to have an entry lock device via which samples can be introduced to the UHV chamber through an airlock. If the surface area of metal exposed to the air is kept very small, then it it possible to introduce or remove a sample without seriously degrading the UHV conditions. This reduces the number of bakeouts required, so saving the whole system from slightly accelerated corrosion processes at the bakeout temperature and avoiding waste of time before the sample can be observed and measured. Often the airlock opens into a subsidiary UHV chamber in which the sample can be cleaned or otherwise prepared before being transferred in UHV to the main measurement chamber. This procedure allows the principal measuring devices (electron guns, photon sources, spectrometers, etc.) to be maintained in very clean conditions, which means that their properties are very stable.

Having placed a sample for study in the vacuum system and having achieved a UHV environment for it, the next step is often to attempt to obtain an atomically clean surface upon which to conduct experiments. Each crystal face of each material presents its own individual cleaning difficulties, and many person-months of effort can be expended in discovering how to produce a clean surface upon which to start study. This is not the place for a detailed summary of techniques and their applicability, but an idea of some of the possible approaches is given in the following list.

1. In some cases, the act of achieving a UHV environment simultaneously cleans up the surface for study. For example, the cleavage face of mica and the (100) cleavage faces of the alkali halides (NaCl, LiF, NaF, KCl, etc.) can be cleaned in this simple way.

2. As will be seen in Chapter 2, common contaminants on many surfaces are oxygen, carbon, and sulphur, which are chemically bound (chemisorbed) to the surface of the material under study. Sometimes these can be removed by heating the sample *in situ*, whereupon the contaminants may desorb into the vacuum as volatile oxides, sulphides, or carbides, or may dissolve into the solid leaving a negligible level of contamination behind. Temperatures near to the melting point of the sample material may be required to realize reasonable rates of removal of these contaminants. An example of this method is the removal of oxygen and carbon from the (111) surface of silicon. On heating for 1 or 2 minutes at 1370 K the surface carbon dissolves into the silicon, leaving behind sub-monolayer levels of contamination.

3. If the chemisorbed contaminants cannot be removed by heating alone, they can sometimes be removed by heating in an atmosphere, which produces a volatile compound of the contaminant. Thus, surface oxides can sometimes be removed by reduction in a hydrogen atmosphere.

4. More stubborn contaminants can be physically knocked off the surface by bombardment with noble gas ions (e.g. Ar^+, Ne^+) (Redhead 1968). This is usually effective in removing contaminants, but is often accompanied by disturbance of the surface atoms of the sample material. For instance, carbon and sulphur contamination can be removed from (100) surfaces of nickel monoxide by bombardment with 200 eV Ar^+ ions, but the nickel monoxide surface is disordered in the bombardment. The surface can be reordered by annealing, but this can result in the reappearance of contaminants by diffusion out of the bulk of the crystal—so-called surface segregation. A combination of bombardment and anneal conditions has to be found empirically which cleans up the surface and allows re-ordering without unacceptable contamination.

5. Some materials cleave naturally on particular crystal faces when struck with a blade in a direction parallel to that face. This property can be exploited to prepare clean surfaces *in situ* and is simple and direct. Examples are (100) alkali halide faces, (111) faces of materials like calcium fluoride, and even some materials like silicon and beryllium, which will cleave at liquid nitrogen temperatures. However, this method is limited to a few faces of a small number of materials and sometimes cleavage steps occur in such a surface. These steps can complicate the interpretation of some kinds of experiment.

6. Evaporation on to a suitable substrate can be used *in situ* to prepare thin films of polycrystalline or single-crystal type, whose clean surfaces can be the subject of subsequent study. The contamination on the substrate can be 'buried' by the deposited film, which may then be acceptable for clean surface studies. The growth of oriented single-crystals by this technique is a process known as epitaxy, and is described in Chapter 6.

The criteria for deciding if a clean surface had been realized in practice used to be based upon the repeatability of an observation between many samples of the same material prepared in the same way. There is little logical basis for this belief, but, at the time, little else could be done. Although this approach has to be resorted to occasionally, it is now more usual to have *in situ* assessment, based upon one of the techniques of electron spectroscopy described in Chapter 2.

2
Surface chemical composition

The first questions to be asked about a surface are: what atoms are present and what are their concentrations? If techniques can be found to provide the answers to these questions, then the next, more detailed, question is: how are the atoms bound to each other? These are the questions of surface chemistry.

The extension of bulk techniques to surface studies

In order to decide whether or not any particular conventional analytical technique will have sufficient sensitivity for useful application to the determination of surface chemical composition, it is first necessary to decide what order of magnitude of mass is to be determined. Taking one atomic monolayer of pure aluminium as a sample it is readily calculated that this face-centred cubic (f.c.c.) material must contain about 2×10^{15} atoms cm^{-2}. Since the mass of each atom is about 4.5×10^{-23} g, a monolayer of aluminium has an areal density of about 10^{-7} g cm^{-2}. A determination of such a monolayer to 1 per cent accuracy therefore requires technique capable of measuring 10^{-9} g for each square centimetre of monolayer available. Using this figure as a yardstick, a number of well-established techniques are compared in Table 2.1.

The analytical methods of bulk chemical analysis have been highly developed to improve their sensitivities, largely because of the demands of thin-film technology and the semiconductor industry (e.g. Maissel and Chang 1970). The microchemical techniques require sufficient sample material to prepare a 1–10 ml solution with approximately 100 μg of sample. Thus, by the yardstick chosen above, 10^3 monolayers cm^{-2} are required. This demand arises from the loss of solvent by evaporation during analysis and by loss of solution during transfer from one container to another.

In colorimetry the sample is dissolved off its substrate, and ions in this solution are reacted with specially selected compounds in order to produce a complex with a distinctive absorption spectrum in the visible or near-ultraviolet. Use is then made of Beer's law which states that, for dilute solutions, the optical absorption is proportional to the concentration of the absorbing ion. The method has to be calibrated by using standard solutions, and reagents are available to produce suitable complexes with most elements in the periodic table.

In flame spectrometry the solution is aspirated into a high-temperature flame and either the emitted radiation is analysed with a spectrometer and the intensity of a selected wavelength is measured or the radiation from a

Table 2.1   Some techniques for chemical analysis in bulk

Technique	Physical basis	Approximate sensitivity (monolayers cm^{-2})	Approximate minimum amount required (g)	Destructive or not	Physical limit to sensitivity or limitations
Solutions					
Volumetry	Titration	10	10^{-4}	D	Reagent purity. Adsorption on laboratory ware surfaces.
Spectro-photometry (colorimetry)	Optical absorption	3	10^{-4}	D	Suitable complexes required. Reagent purity.
Polaro-graphy	Current–voltage relations	10	10^{-5}	D	As volumetry.
Flame spectro-metry	Emission spectrum	10^{-1}	10^{-4}	D	Optical detector sensitivity. Noise in fluctuating emission.
Solids					
Mass spectro-scopy	Mass to charge ratio (m/e)	10^{-6}	10^{-12}	D	Ion detector sensitivity.
X-ray emission	Stimula-tion of character-istic X-rays	10	10^{-4}	ND	X-ray detector sensitivity. Background of 'white' X-rays.

discharge tube containing the element to be determined is absorbed in the flame and the amount of absorption is measured. Again the method is calibrated against standards, and sensitivities of sub-monolayer order are possible.

If it is desirable not to dissolve the sample away then techniques using the solid are required. The most common are mass spectroscopy and X-ray emission spectroscopy, the former being overwhelmingly the most sensitive of established chemical methods. For thin-film and surface applications of mass spectroscopy, atoms are desorbed from a surface thermally by heating the solid near to or slightly above its melting point or by bombarding the surface with elelctrons, atoms, or ions. The ejected atoms pass into the mass spectrometer, where they are first ionized by electron bombardment. This process can be carried out with very high efficiencies and the ions thus generated passed into a mass filter, which, for particular combinations of magnetic and/or electric fields, allows only ions of a particular mass to charge ratio (m/e) to reach a detector. If the detector is an electron multiplier, then one ion reaching it can result in about 10^6 electrons arriving at its collector electrode. It is the combination of high efficiency of ionization, good resolution of m/e and high gain in the detector that results in the high performance of mass spectrometers and their widespread application. The disadvantages of using mass-spectroscopic techniques are associated with controlling the desorption of atoms from the surface when heating the sample or bombarding it with ions.

Perhaps the most widely used mass spectrometer is the quadrupole type illustrated in Fig. 2.1 (e.g. Redhead 1968). Here, ions generated at position S by electrons from the filament F pass through an aperture A into a region

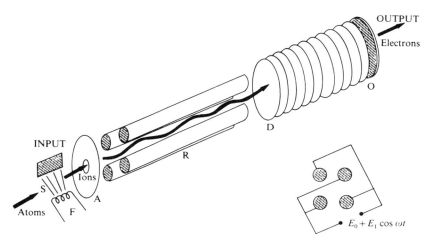

Fig. 2.1 The quadrupole mass spectrometer. For particular values of E_0, E_1, and ω, ions of particular m/e pass between the rods on a stable oscillating path. The inset indicates how the rods are electrically connected.

containing four accurately parallel round conducting rods R. The electric field E in the region between the rods has a steady component E_0 and a component $E_1 \cos\omega t$ oscillating at radio frequencies ω. For given geometry and particular values of E_0, E_1, and ω, only ions of a particular m/e can oscillate in stable orbits and reach the detector surface D, which is the first dynode of an electron multiplier. Electrons generated by the arriving ion are accelerated down the multiplier, generating more electrons as they strike successive dynodes. The signal corresponding to the particular m/e value finally emerges as a pulse of electrons at O for each ion arriving at D. If the rate of arrival of ions at D is sufficiently high the signal at O is a steady current which can be measured with a sensitive amplifier. The mass spectrum can be explored by sweeping through a range of frequencies ω, and the resolving power $m/\Delta m$, where m is the smallest observable mass difference, can be varied by altering the ratio of E_1 to E_0.

In X-ray emission spectroscopy the solid sample is bombarded with electrons or high-energy X-rays so that many atomic levels in the solid are ionized. The atoms return to their ground states by emitting characteristic X-rays. The emitted photons pass into an X-ray spectrometer and the intensity of a selected X-ray line is measured. This is the only non-destructive technique in Table 2.1, but the cross-sections for X-ray emission can be rather small and the efficiency of X-ray spectrometers is low so that the sensitivity of the technique is not normally high enough for analysis of sub-monolayer quantities, except under the most favourable circumstances. By illuminating the surface with a grazing incidence electron beam and detecting X-rays in grazing emission it is possible to reach sensitivities for the surface atoms comparable to those attainable with Auger electron spectroscopy (see later). If electrons are incident upon the sample and the detector is sensitive to the energy of the emitted X-rays then the technique is called energy dispersive X-ray spectroscopy (EDX). A common detector for this kind of spectroscopy is a piece of lithium-doped silicon, within which the X-rays create a number of electron-hole pairs proportional to their energy. If, on the other hand, the detector is a spectrometer sensitive to the wavelength of the X-rays then the technique is naturally called wavelength dispersive X-ray spectroscopy, or WDX.

Specifically surface techniques

The destructive character of most microchemical methods and their relative insensitivity in the context of surface studies leads to a search for other methods. The only exception in Table 2.1 is mass spectroscopy, which, in spite of its destructive character, is so sensitive that it is of great importance in surface physics, and its application in studies of adsorption and the interaction of atoms or molecules with surfaces is described in Chapter 6. The sensitivity of mass spectrometry is also used in a method called secondary ion mass spectroscopy (or SIMS) which is described on p. 37.

The most important methods of surface chemical analysis involve energy analysis of electrons emitted from a surface after it has been bombarded with ultraviolet photons, X-ray photons, or electrons. All these methods of electron spectroscopy use the fact that some of the electrons emitted have energies characteristic of particular combinations of the energy levels in the solid and so are characteristic of the types of atoms contained in the solid. The process involved in ultraviolet photoelectron spectroscopy (UPS), X-ray photoelectron spectroscopy (XPS), and Auger electron spectroscopy (AES) are described using one-electron energy level diagrams in Fig. 2.2.

Many techniques are used for the energy analysis of the electrons leaving the specimen. Four common methods are illustrated in Fig. 2.3 and are described in more detail by Sevier (1972). The two most important characteristics of an electron energy analyser are its energy resolving power and the fractional solid angle Ω of electrons which it will accept for analysis. The resolving power ρ is determined by the slit widths and geometrical factors of the design. It is related to the energy of the electrons E and the spread of energies ΔE_A passed by the analyser through the equation

$$\rho = \frac{E}{\Delta E_A}.$$
(2.1)

Most systems operate at constant resolving power as the analyser is swept through the range of energies under study. The sensitivity of the analyser is determined by the fraction $\Omega/2\pi$, since 2π is the total solid angle for backscattering. In most analysers, increased sensitivity can be obtained only at the expense of reduced resolving power.

The highest-precision photoelectron instruments were set up and operated by Siegbahn and his co-workers (1967), but most of these suffered from the disadvantage of being physically large conventional vacuum instruments dedicated to electron spectrometry only. An exception is a very powerful UHV electron spectrometer described by Gelius *et al.* (1990). This has a large hemispherical analyser of the general type indicated in Fig. 2.3(d) with a more complex input lens stack than is shown there. It is also equipped with an X-ray monochromator and an array of detectors capable of detecting electrons at several different energies simultaneously. This instrument has been reported as achieving an energy resolution of 0.27 eV, a high collection efficiency, and the ability to record XPS data from an area as small as 23 μm diameter. Of course, such high performance does not come cheaply!

One kind of very simple UHV compatible analyser is that of Fig. 2.3(b) which is useful because it requires the same electron optics as that used in low-energy electron diffraction LEED (Chapter 3), and is therefore more flexible in its applications. In general, the retarding potential analysers are used for identification of what elements are present upon a surface and the more powerful types of Fig. 2.3(a) and (d) for detailed study of the electron spectrum. The cylindrical mirror analyser of Fig. 2.3(b) has good resolving power and high collection efficiency combined with reasonable physical size and is useful for fast collection of results. It is in widespread use.

Spectrometers of the type shown in Fig. 2.3(a) and (d) can be adapted to detect electrons simultaneously in several different adjacent energy ranges.

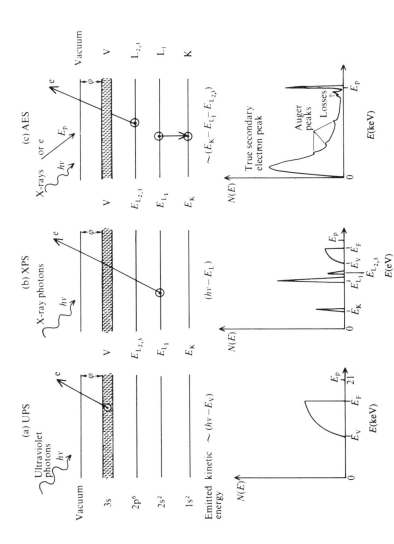

Fig. 2.2 Three kinds of electron spectroscopy. The one-electron energy level diagrams at the top are for a mythical ideal free electron metal. They are *not* drawn to scale. At the extreme left are the electronic configurations in which the letters s, p, d, f, ... signify electrons having orbital (angular momentum) quantum numbers 0, 1, 2, 3, ... ; the numbers to the left of the letter denote the principal quantum number of one orbit; the superscript to the right of the letter denotes the number of electrons in the orbit. At the extreme right are the labels for the orbitals used in X-ray spectroscopy. The levels E_K, etc. correspond to the binding energies of the electrons measured from the vacuum level. The graphs at the bottom indicate the number $N(E)$ of electrons with kinetic energies between E and $E + dE$ which will be measured in the electron energy distribution leaving the solid. The horizontal scales correspond approximately to those that might be found in common practical cases.

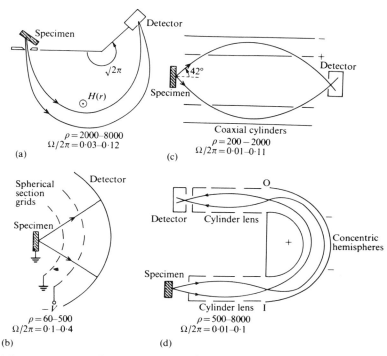

Fig. 2.3 Four methods of electron energy analysis after photoemission or an Auger process. Typically accessible ranges of ρ and $\Omega/2\pi$ are indicated for each method. (a) The magnetic double-focusing spectrometer. Here the field H is made to be proportional to $r^{-1/2}$ for an electron trajectory of radius r. (b) The electrostatic retarding potential analyser. The potential $-V$ on the second grid results in electrons with kinetic energies greater than eV reaching the detector. The grids and detector are concentric spherical sections with their centres on the sampling surface and at the centre of the region of electron emission. (c) The electrostatic cylindrical mirror analyser (CMA). Appropriate potentials on the outer and inner cylinders result in electrons with a particular kinetic energy being focused at the detector aperture. (d) The electrostatic concentric hemispherical analyser (CHA). The electron source on the sample surface is focused on the entrance aperture I of the analyser. With appropriate potentials on the inner and outer hemispheres electrons with a particular kinetic energy are focused on the output aperture O. The three-element output lens then focuses this image on to the (real) aperture of the detector.

This is because they spread electrons with different kinetic energies over a plane in space at the output end of the spectrometer. Several solid state detectors can be placed side by side in this plane. The detectors in this kind of multi-channel spectrometer are often electron multipliers in the form of thin micro-channel plates. The advantage of using a multi-channel spectrometer is that it takes less time to acquire the spectrum over a given energy range than it does with a single detector. The disadvantages are increased complexity and cost.

Photoelectron spectroscopy

The photoelectron spectra of very large numbers of atoms and molecules have been studied intensively by Siegbahn and his co-workers (1967), and this work provides an important base upon which to start investigation of solids and solid surfaces. Following Siegbahn, XPS is very often referred to as electron spectroscopy for chemical analysis (ESCA). The whole of this field has been reviewed by Brundle and Baker (1981).

In both types of photoelectron spectroscopy, if the incident photon has sufficient energy hv it is able to ionize an electronic shell and an electron which was bound to the solid with energy E_B is ejected into the vacuum with kinetic energy E_k. By conservation of energy,

$$E_k = hv - E_B, \qquad (2.2)$$

neglecting the very small recoil energy of the emitting atom. If the incident radiation is monochromatic and of known energy and if E_k can be measured using an electron energy analyser, then the binding energy E_B can be deduced. In UPS the source of radiation is often a helium discharge lamp, which can be made so as to operate at 21.2 eV and at 40.8 eV. In XPS the radiation is usually obtained from X-ray tubes with aluminium or magnesium anodes, which give lines at 1487 eV and 1254 eV, respectively. Alternatively, X-rays may be used from a synchrotron radiation source (see later) which has the advantages of being very bright and, in combination with a monochromator, can be tuned to the X-ray energy of interest. The helium radiation has insufficient energy to eject deep core electrons and is used mostly to study the electrons in the valence band of the solid. On the other hand, the X-rays usually have sufficient energy for ionization of core levels in many elements and can be used to study electrons in both band and core states.

The width of a feature observed in a photoelectron spectrum will depend upon the intrinsic width of the level from which the photoelectron is ejected, the width of the incident radiation (because hv appears in eqn (2.2)), and the window ΔE_A of the analyser. In XPS the width of the K_α radiation† incident upon the sample is usually the controlling instrumental factor and this is about 1–1.4 eV unless an X-ray monochromator is incorporated. In UPS the anlayser window is normally the controlling parameter as the width of the radiation from a helium II‡ resonance source is close to 1 meV.

Because the number of photons arriving in the incident beam from a laboratory source (as opposed to a synchrotron) can often be rather small, the intensity of photoelectron lines may be low and individual electrons may have to be counted as they arrive at the detector.

---

†In X-ray notation, $K_{\alpha1}$ is the line in the X-ray emission due to an electron in the L_3 sub-shell falling into the K-shell hole; $K_{\alpha2}$ is the line due to an electron in the L_2 sub-shell falling into the K-shell hole. $K_{\alpha3,4}$ arises from doubly ionized K-shell initial states.

‡In this spectroscopic notation, He II is the radiation emitted by He^+ ions in the transition from their first excited state to their ground state.

Part of the XPS spectrum obtained from a silicon niobium oxide using monochromatic Al K_α incident X-rays is reproduced in Fig. 2.4. This figure shows the counting rates required, on the vertical scale, and demonstrates clearly the narrow core-like levels labelled 1s, 2s, and 2p etc. The peaks can also be labelled by conventional X-ray notation, where the letters K, L, M, N, O, . . . , etc. are assigned to orbitals with principal quantum numbers $n = 1, 2, 3, 4, 5, . . . ,$ etc. The suffixes give the orbital and total angular momenta quantum numbers l and j of the hole left after photoemission according to the convention in Table 2.2. The X-ray notation is related to the spectroscopic notation in columns 4 and 6 of Table 2.2. In addition, an oxygen Auger peak occurs in Fig. 2.4(a)—the O 1s XPS peak arises by photo-ionization of the oxygen 1s level and the oxygen Auger peak results when the 1s core hole is filled by a valence band electron and a second valence band is emitted. The relative energies and intensities of the photoelectron and the corresponding Auger peaks occurring in the same XPS spectrum can be valuable source of information about the electronic processes going on in a solid when an inner shell hole is formed.

Figure 2.4(b) shows the valence band spectrum of silver obtained with the high resolution spectrometer described by Gelius *et al.* (1990). The broader region between binding energies of about 3 to 8 eV is due to the silver 4d electrons and contains interesting fine structure which can only be observed with a monochromatic X-ray source. The step near zero binding energy is due to the Fermi edge in a weak s-p band.

First-principles theoretical calculations of the energies of photoelectron lines from atoms can be made under two extreme kinds of approximation, illustrated by the configurational diagram of Fig. 2.5. In the '*sudden*' *approximation* the wave functions are assumed to be unchanged during the time that the photoelectron is emitted (and the repulsive force on the outer electrons due to the coulomb interaction with the inner electron is removed) and the binding energy of the photoelectron is just the difference E_{BS} in Fig. 2.5. In the *adiabatic approximation* the wave functions relax to their ionic form before the photoelectron has left the region of the atom and the calculated binding energy will correspond to E_{BA} in Fig. 2.5. The discrepancy between E_{BA} and E_{BS} for core levels is found to be greater than the difference between the best theoretical value and observation. E_{BA} is the best fit to observation, always falling 1–12 eV above the observed binding energy for K-shell calculations on elements with atomic numbers Z between 6 and 23, and for which E_K varies from 288 eV to 5469 eV. This difference may be due to the fact that the theory is for an isolated atom of the material and the observation is usually on atoms incorporated in a solid.

The valence band spectrum of gold is shown in Fig. 2.6, which was obtained by Spicer (1970) using UPS with a set of helium radiations having 5.4 eV $< h\nu <$ 21.2 eV. Although the main features of the valence band structure observed by XPS or UPS do agree, there is considerably more structure in the

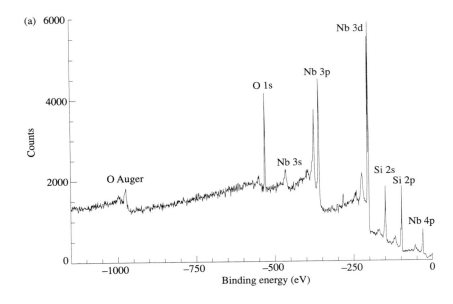

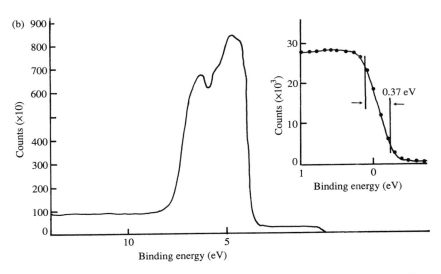

Fig. 2.4 (a) XPS spectrum of the outer levels of solid silicon niobium oxide. The incident X-rays are monochromatized Al K$_\alpha$. The binding energy scale has its zero on the right and is set at the Fermi level. (By courtesy of J. A. D. Matthew, 1993.) (b) Valence band XPS data from silver using monochromatic X-rays and the high performance spectrometer described by Gelius *et al.* (1990). The Fermi edge is shown in the inset where the instrumental energy resolution of 0.37 eV is marked.

Table 2.2 Atomic and X-ray notations

Quantum numbers			Atomic notation	X-ray suffix	X-ray notation
n	l	j			
1	0	1/2	1s	1	K
2	0	1/2	2s	1	L_1
2	1	1/2	$2p_{1/2}$	2	
2	1	3/2	$2p_{3/2}$	3	$L_{2,3}$
3	0	1/2	3s	1	M_1
3	1	1/2	$3p_{1/2}$	2	
3	1	3/2	$3p_{3/2}$	3	$M_{2,3}$
3	2	3/2	$3d_{3/2}$	4	
3	2	5/2	$3d_{5/2}$	5	$M_{4,5}$
4	0	1/2	4s	1	N_1
4	1	1/2	$4p_{1/2}$	2	
4	1	3/2	$4p_{3/2}$	3	$N_{2,3}$
4	2	3/2	$4d_{3/2}$	4	
4	2	5/2	$4f_{5/2}$	5	$N_{4,5}$
4	3	5/2	$4f_{5/2}$	6	
4	3	7/2	$4f_{7/2}$	7	$N_{6,7}$

See e.g. Kuhn 1969.

data obtained by UPS. This is not only because of the greater energy resolution of the UPS experiment (ΔE in the UPS experiment is about 0.1 eV, but in the XPS experiment it is about 1 eV) but also because of differences in the two processes. The interpretation of the UPS spectra is complicated (Spicer 1970; Fadley and Shirley 1970) by the following:

1. The rapid variation of the mean free path for electron–electron scattering (Fig. 2.7) at low kinetic energies. Thus, for fixed photon energy, electrons originating from the bottom of the conduction band have a greater mean free path than those originating from near the Fermi level.

2. If an electron is ejected with low kinetic energy the probability of photoemission will be affected by the density of states available to the

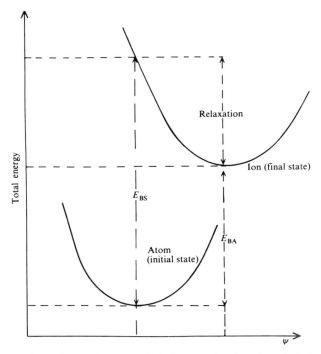

Fig. 2.5 Comparison of the sudden and adiabatic approximations for calculating the binding energy. The total energy of the atom and the ion is shown as a function of a measure of the spatial extent ψ of the wavefunction of the remaining electron.

photoelectron. A variation of this empty density of states with energy will affect the shape of the observed $N(E)$ curve for the photoelectrons.

The surface sensitivity of the photoelectron methods does not depend upon the penetrating depth of the incident radiation but rather upon the probability that a photoelectron, once generated, will be able to escape to the surface without further energy loss. Electrons can lose energy in a number of ways listed below, each of which is a quantized process.

1. The smallest energy losses, of the order of a few tens of meV, are due to the excitation of lattice vibrations or phonons. These losses are so small that they are usually detected only in experiments using high-resolution spectrometers. In other cases they are normally part of the low-energy side of a photoelectron peak.

2. Electron–electron interactions can excite collective density fluctuations in the electron gas in a solid. These fluctuations, known as plasmons, are quantized with energies in the range 5–25 eV, the energy depending upon the

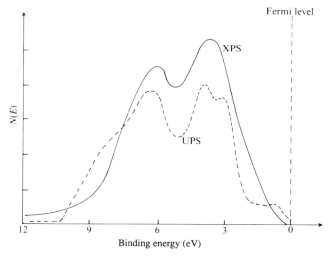

Fig. 2.6 The one-electron density of states for gold obtained by UPS (Spicer 1970) compared with that obtained by Siegbahn *et al.* (1967).

density of the electron gas and whether or not they are three-dimensional fluctuations—*bulk plasmons*—or two-dimensional fluctuations at the surface—*surface plasmons*. The subject is reviewed by Kittel (1986). The energy ΔE_B lost to a bulk plasmon is given by

$$\Delta E_B = n \left(\frac{ne^2}{\varepsilon_0 m} \right)^{1/2} \tag{2.3}$$

in a free-electron gas of density n. The energy ΔE_S lost to a surface plasmon is given by:

$$\Delta E_S = \left(\frac{\Delta E_B}{\sqrt{2}} \right) \tag{2.4}$$

3. Various single- and double-particle excitations can occur. Thus an electron may lose energy by raising a second electron from its ground state to an empty state in the solid (e.g. interband transition or a transition from a core level to an impurity state). Alternatively, it may ionize a level with the ejection of another photoelectron, perhaps leading to the generation of another Auger electron (p. 25).

The net effect of these inelastic processes is difficult to calculate, but it is known that in metals and semiconductors the mean path of an electron prior to inelastic scattering will vary with its kinetic energy. At very low energies the electron will be unable to excite any of the above losses and the mean free path

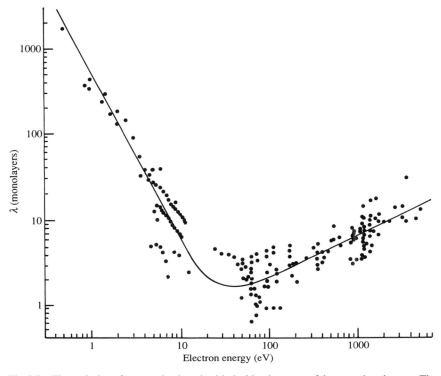

Fig. 2.7   The variation of attenuation length with the kinetic energy of the emerging electron. The dots correspond to the attenuation lengths measured for electrons of the kinetic energy given by the scale on the abscissa for a wide variety of photoelectron and Auger peaks in many different materials.

Universal curves of this kind should be treated only as a guide to the attenuation length of the escaping electron. As described in the text, there are several mechanisms responsible for the losses causing this escape depth, and there is no reason to expect that they will add up in such a way that universal curves can always be drawn. (After Briggs and Seah 1990.)

will be long. At very high energies, the cross-sections for exciting the losses fall, and again the mean free path will be long.

The experimental measurement of the escape distance for an electron does not lead to a quantity directly comparable with the inelastic mean free path. This is because experiments measure the attenuation of the number of electrons escaping to be analysed as the thickness of an overlayer is varied. This attenuation includes both inelastic and elastic scattering processes. The subject is discussed carefully by Tanuma *et al.* (1990), who present a method for calculating the inelastic mean free path using a knowledge of the experimental optical properties. As indicated in Fig. 2.7, the attenuation length is believed to pass through a minimum below about 1 nm deep at energy near 40 eV. Even at 1000 eV the attenuation length in metals is probably less than about 2 nm (about 10 atomic monolayers).

Thus the surface sensitivity of the photoelectron methods will vary with the kinetic energy of the ejected electron, but is highest in the range 50–200 eV. In UPS using 21 eV radiation the depth sampled will be greater than in using UPS at 41 eV because of the low kinetic energies in the first case. In XPS the surface sensitivity will vary more extremely because of the wide range of kinetic energies possible. Siegbahn *et al.* (1967) were able to detect and measure 10^{-8} g of iodine in the $3d_{5/2}$ peak at 620 eV—a figure which corresponds to a sub-monolayer coverage in the terms of Table 2.1 (p. 12).

In summary, the techniques of photoelectron spectroscopy can be used to identify the atoms at surfaces by comparing the lines observed with either calculated core level binding energies or experimentally derived spectra from standards. Quantitative assessments of the amount of each element present are more difficult and discussion of this is deferred until pp. 35–7. Information about bonding of surface species can be obtained by either UPS or XPS by studying the change of shape of valence band photoelectron features as the surface species is added. Alternatively, the change in energy of a core level (the chemical shift) as the environment is changed can give bonding information.

Auger electron spectroscopy (AES)

As indicated in Fig. 2.2(c), the Auger process is an alternative to X-ray emission and occurs after an atomic level has been ionized by incident photons or electrons. The hole in the inner shell is filled by one electron from a less tightly bound level and a second electron escapes into the vacuum with the remaining kinetic energy. The energy of this Auger electron is very roughly (see later)

$$E \sim E_K - E_{L1} - E_{L2,3} \qquad (2.5)$$

for the transition shown in Fig. 2.2(c), because $E_K - E_{L1}$ is the amount of energy released by an electron falling from the L_1 shell to the initial state hole in the K shell, and the electron escaping uses up $E_{L2,3}$ of this amount to overcome its own binding energy. Again, then, the emitted electrons due to this process will have energies characteristic of the levels of the atoms whence they came and energy analysis will enable identification of the materials present. The electron described by the process of eqn (2.5) would be referred to as a $KL_1L_{2,3}$ Auger electron. If one or both of the final state holes is in the conduction band of a metallic sample than it is conventional to use the notation V for each of these holes. Thus, an $L_{2,3}VV$ Auger transition would involve an initial state hole in the $L_{2,3}$ shell and two final state holes in the conduction band.

For light elements (atomic number $Z < 20$) Auger emission is more probable than X-ray emission for a K-shell initial state hole, and for $Z > 15$ it is almost the exclusive process. For higher Z, Auger processes dominate for initial state holes in other shells. Thus, if the primary electron beam has an energy below 1000 eV Auger processes will predominate. These high probabilities of Auger emission, coupled with the high flux of incident electrons, which is easily achieved in practice, lead to Auger electron spectroscopy (AES) being an extremely sensitive technique for surface chemical analysis which is in widespread use.

All the forms of spectrometer shown in Fig. 2.3 are used for AES. To obtain highest sensitivity the incident beam is of electrons arranged to arrive at the surface near grazing incidence by the argument shown in the caption to Fig. 2.8. In order to obtain numerical estimates of the current of Auger electrons that will be collected in an experiment, Bishop and Riviere (1969) have shown that the equation in this caption has to be modified as follows.

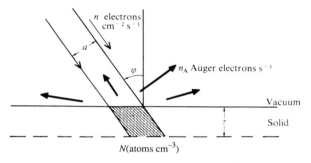

Fig. 2.8 The Auger current generated by a beam current I_o within an escape depth τ.

Assume: Incident beam has cross-sectional area A. One incident electron causes Y Auger electrons per atom. All Auger electrons generated within τ of surface escape—Auger electrons generated in the rest of the solid suffer losses in escaping and do not contribute. N atoms cm^{-3} contribute to the Auger process in question.

Then: Volume corresponding to hatched area above contains all atoms contributing to Auger signal. Number of contributing atoms $= N\tau A \sec \phi$; nA incident electrons per second give $nA Y N\tau A \sec \phi$ Auger electrons per second. Auger current $I_A = I_o N\tau A \sec \phi$ for incident current I_o. But YA is the cross-section Φ for Auger emission back into the vacuum, so
$$I_A = I_o N\tau \Phi \sec \phi.$$

1. The incident electron current I_o ionizes atoms on the surface directly, but also causes secondary electron emission from underlying atoms. Some of these secondary electrons will also have enough energy to ionize the surface atoms. These energetic secondary electrons are often referred to as backscattered electrons. Ionization by these electrons can be allowed for by incorporating with I_o a factor r, the *Auger backscattering factor*, which increases the effective value of I_o.

2. If the cross-section for ionization by an incident electron is Φ then the

only means by which the ion can decay to its ground state are by emission of a characteristic X-ray or by emission of an Auger electron. If the proportion of decays attributable to X-ray fluorescence is ω (the *fluorescence yield*) then the Auger cross-section Y_A must be simply $(1-\omega)\Phi$.

3. In any practical apparatus, all the Auger current I_A emitted into 4π sr will not be collected but some fraction $\Omega/4\pi$ will arrive at the detector, where Ω is the solid angle accepted by the analyser. This is valid if it is assumed (reasonably) that Auger emission is isotropic.

The net effect of these modifications is that the total Auger current observed, I_A, will be given by

$$I_A = N\Omega I_o \tau r(1-\omega)\Phi \sec \phi/4\pi. \tag{2.6}$$

Reasonable values for the parameters in eqn (2.6) lead to values of I_A between 10^{-14} and 10^{12} A, depending upon N, I_o, Ω, and the energy E_p of the incident electrons (which affects Φ). As indicated in Fig. 2.2(c), this current is superimposed upon a backgound current due to secondary electron emission which carries no direct information about the type of atoms present and which may be 5 or 10 times larger than I_A. Special electronic techniques can be devised to present clearly observable Auger signals in the presence of this background, but they result in the observation of the differential $d(N(E)/dE$ (or $N'(E)$) of the energy spectrum instead of $N(E)$, which is shown in Fig. 2.2. However, it is now most common practice to present spectra in the direct $N(E)$ form because this reveals the shape of the Auger peaks directly and has better signal to noise ratio performance than the $N'(E)$ form.

An Auger spectrum in the $N(E)$ form, obtained using a concentric hemispherical analyser of the general type shown in Fig. 2.3(d), is given for two Si(111) surfaces in Fig. 2.9. After careful chemical cleaning in the laboratory the silicon specimen is placed in the spectrometer, and its Auger spectrum is as indicated in Fig. 2.9(a) and (c). By using either tables of atomic energy levels (Bearden and Burr 1967) and relation (2.5) or spectra from known materials, the features can be assigned to C, O, and Si.

This is a common type of result, the contaminants C and O being present on many kinds of surfaces chemically cleaned at atmospheric pressure. After cleaning in the UHV chamber by heating near the melting point or bombardment with noble gas ions the contaminants can be removed from the silicon surface and the 'clean' surface spectrum of silicon obtained (Fig. 2.9(b)). Careful use of eqn (2.6) indicates that the residual carbon signal in Fig. 2.9(b) and (d) corresponds to less than 5 per cent of an atomic monolayer of carbon. This demonstrates the high sensitivity of Auger electron spectroscopy for surface species—a property which arises because of the high ionization efficiency of the incident electrons and the small mean free paths for inelastic scattering of the comparatively low-energy Auger electrons.

If a high resolving power electron-energy analyser is used (e.g. that of Fig.

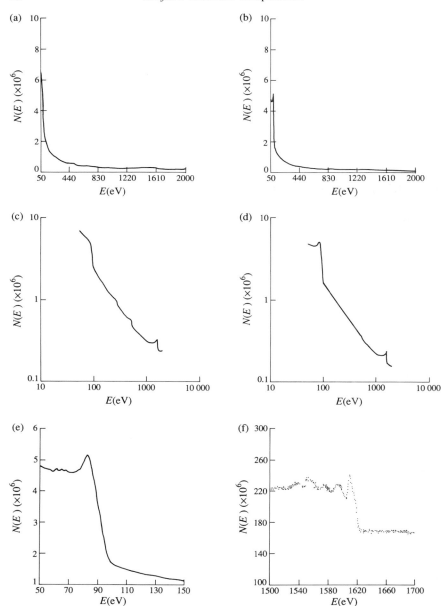

Fig. 2.9 *N(E)* Auger spectra from Si(111). (a) The chemically cleaned surface as inserted in the UHV system. The spectrum is presented as *N(E)* versus *E*. The LVV feature of Si near 90 eV is discernible, but is not a clear peak. Small peaks due to C (255 eV) and O (505 eV) are visible. A very small feature due to a Si KLL process is just visible near 1610 eV. (b) The 'clean' surface Auger spectrum showing only Si features. The Si LVV peak is now clearer and the peaks due to C and O have disappeared. (c) A log–log presentation of the spectrum in (a). This kind of plot compresses

2.3(d) combined with a high retardation of the electron velocities through the input cylinder lens) then fine structure can be observed in the Auger spectra of some materials. This had been seen many years ago for gaseous samples (Siegbahn 1967) and very similar effects are seen in solids. The $N(E)$ curves for the KLL Auger lines of oxygen and magnesium in the (100) surface of megnesium oxide obtained with an instrument with resolving power of 1200 are shown in Fig. 2.10. The fine structure can be explained in terms of L–S coupling of two final state holes left in an atom near the surface after the emission of an Auger electron (for an account of L–S coupling see e.g. Kuhn 1969). Many solids show these quasi-atomic Auger spectra, the features being different from the gas only in that they are all shifted in energy and greater in linewidth. Even higher resolution Auger spectra have been reported, for example, by Weightman (1982) who has been able to show how careful measurements of both AES and XPS peak positions allows the isolation of initial and final state contributions, which can then be related to the results of other experimental techniques.

Careful examination of the energies of observed Auger lines shows that relation (2.5) is not accurately obeyed. The correct energy of an Auger line is not given by differences of the three binding energies as obtained from XPS. This is because of the interactions between the two final state holes. Referring, as an example, to Fig. 2.2(c), once the L_1 electron has filled the initial state hole in the K shell, then the binding energy of the $L_{2,3}$ electron is increased because the Coulomb repulsion of an L_1 electron has been removed. This hole–hole interaction in the final state configuration will depend upon whether the holes are both in core shells, one in a core shell and one in a band, or both in a band. Empirical techniques have been devised to allow for this effect, and these and other complications are discussed by Carlson (1975) and by Madden (1981). Accuracies of about 5 eV can be obtained by using eqn (2.7) for an Auger transition involving the three levels A, B, and C.

If the two final state holes are both in the conduction band the electron gas

---

Fig. 2.9 *continued*

the vertical scale, so making the peaks more discernible as well as revealing linear regions between the peaks. This linearity in a log–log presentation means that the background between the peaks is following a power law of the kinetic energy E. (d) The log–log presentation of (b) showing the clear Si LVV and KLL features. (e) An expanded region of the LVV region of the clean Si spectrum in (d). A plasmon loss peak can be seen about 16 eV below the main peak. (f) An expanded region around the KLL peak of Si in (d). Fine structure is clear in this part of the spectrum.

All of these data were obtained in a scanning Auger electron microscope (see later). The primary energy was about 20 keV in a beam carrying about 5 nA of current focused to a spot on the sample of about 200 nm diameter. The angle of incidence was 30°. Electrons emitted at 32° to the surface normal entered a concentric hemispherical analyser which collected about 2 per cent of the electrons emitted by the sample.

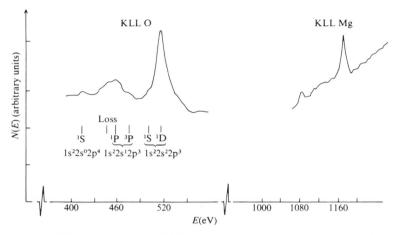

Fig. 2.10 The KLL spectra from MgO(100) obtained using a concentric hemispherical analyser with revolving power 1200. The fine structure is assigned using L-S coupling of the two final-state holes. The final state of the atom (after Auger emission) is indicated using the atomic notation of Table 2.2, the superscripts indicating the number of electrons in each orbital. (After Bassett *et al.* 1972.)

screens the Coulomb interaction between them, and so eqn (2.5) gives a reasonable approximation to energies of AVV Auger features. If both final state holes are not in the conduction band, then an approximate expression for the kinetic energy of an Auger electron is given by eqn (2.7). In this equation, E_A, E_B, and E_C are the binding energies

$$E_{ABC}(Z) \approx E_A(Z) - \tfrac{1}{2}\{E_B(Z) + E_B(Z+1) + E_C(Z) + E_C(Z+1)\} \qquad (2.7)$$

of electrons in levels A, B, and C. The use of the binding energies for combinations of the atomic numbers Z and $(Z+1)$ as indicated makes crude allowance for the hole–hole interaction. If the peak shapes and satellites are examined closely, then even eqn (2.7) turns out to be an oversimplification. As described in the review by Weightman (1982), Cini (1978) showed that the Auger spectra of metals may be band-like or free atom-like depending upon the size of the interaction between two final state holes on the same atom as compared to the energy width of the conduction band. If the interaction energy is large compared with the bandwidth, then quasi-atomic behaviour is to be expected—the Auger spectra of the solid appear to be broadened and shifted versions of those of the isolated atoms (i.e. atoms in the gas phase). On the other hand, if the interaction is small compared with the bandwidth then the shape of a core valence–valence Auger peak is more like a self-convolution of the density of occupied states in the conduction band—a true solid state profile.

Since both photoelectrons and Auger electrons can be generated when a photon beam ionizes a core level in an atom, both kinds of peak are observed in the photoelectron spectrum (see Fig. 2.4(a)). The differences in the kinetic energies of Auger and photoelectron lines based upon the same level being ionized have been found to be related to the chemical environment of the ionized atom. Since this difference between two energies is measured in the same spectrum it is independent of the choice of the zero of energy—the reference level—and of charging effects. Thus it is a measure of chemical state effects and is independent of the spectrometer properties and of the insulating, semiconducting or metallic properties of the sample. The *Auger parameter*, α, was introduced by Wagner as a measure of this energy difference and is defined as

$$\alpha = E_k(jkl) - E_k(i) + h\nu = E_k(jkl) + E_B(i). \tag{2.8}$$

In these equations $E_k(jkl)$ is the measured kinetic of the photon-excited Auger peak based upon ionization of shell k and $E_k(i)$ is the kinetic energy of the photoelectron line based upon ionization of the same shell. $E_B(i)$ is the binding energy of the same photoelectron line. Compilations of Auger parameter data are particularly useful when presented in the form of graphs with the Auger kinetic energy plotted on the ordinate and the photoelectron binding energies plotted on the abscissa. If the binding energies are ordered in decreasing value to the right, then the variations in α appear on lines with slope 1. Each chemical state appears as a unique point on this graph. Measurements of α have been used to study the coordination of silicon in a variety of silicates, the oxidation of Fe–Si alloys, and charge transfer and core–hole screening effects in CdTe. Waddington and Wagner have detailed some of the explanation and measurements of the Auger parameter in two separate appendices in the book by Briggs and Seah (1990). The use of the graphs of Auger kinetic energies versus photoelectron binding energies has been reviewed by Wagner and Joshi (1988).

The surface sensitivity of Auger spectroscopy can be measured and eqn (2.6) tested by exploiting the observation that some materials can be grown upon others in such a way that the atoms build up in large monolayer-thick islands layer by layer. That this occurs can be established by some of the techniques described in Chapter 3. An example of such growth is found when silver grows upon the (100) surface of clean nickel, and Fig. 2.11 shows the results of an Auger spectroscopy experiment using this system. The silver is deposited at a constant rate and the size of the 355 eV silver and 60 eV nickel Auger peaks measured as a function of time. A clear break in slope is visible when the first monolayer of silver has formed, and it can be seen that 10 per cent of a monolayer of silver could be observed by this method. The experiment can also be used to estimate the mean escape depths of different Auger electrons and these are found to be 0.54 nm for the 355 eV silver electrons and ~ 0.2 nm for the nickel 60 eV electrons.

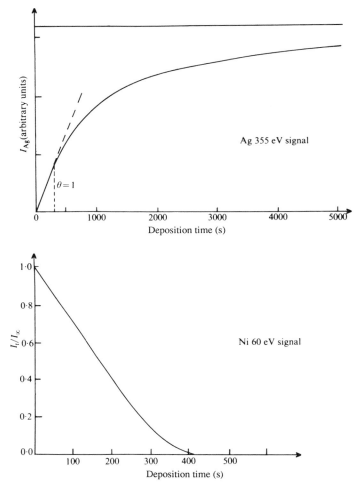

Fig. 2.11 The growth of a silver MNN Auger signal at 355 eV and decay of a nickel MVV Auger signal at 60 eV as silver is deposited at constant rate upon Ni(100). The silver grows up monolayer by monolayer. The break point in the rising curve for silver occurs at a coverage of one monolayer, because at greater coverages the outermost silver atoms shield the first monolayer. (After Jackson *et al.* 1973.)

Scanning Auger microscopy (SAM)

The scanning electron microscope, Fig. 2.12(a), has been well established as an instrument which images a variety of physical properties at or near the surface of a solid (Wells *et al.* 1974). By scanning a focused electron beam across the specimen surface in a TV-like raster while a synchronous raster is scanned

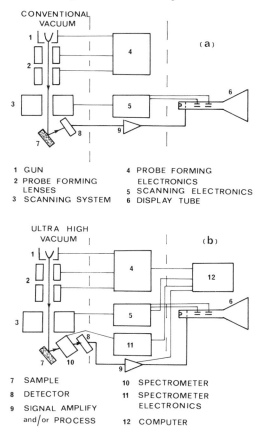

1 GUN
2 PROBE FORMING
  LENSES
3 SCANNING SYSTEM

4 PROBE FORMING
  ELECTRONICS
5 SCANNING ELECTRONICS
6 DISPLAY TUBE

7 SAMPLE
8 DETECTOR
9 SIGNAL AMPLIFY
  and/or PROCESS

10 SPECTROMETER
11 SPECTROMETER
   ELECTRONICS
12 COMPUTER

Fig. 2.12 (a) The principal components in a scanning electron microscope (SEM). (b) Modifications to a conventional SEM to allow scanning Auger electron microscopy. The vacuum becomes ultra-high vacuum and a spectrometer is added to energy analyse the emitted electrons. Computer control of the instrument is common.

across the face of a display oscillograph, a one-to-one correspondence can be set up in the positions of the probe upon the specimen and the beam upon the oscillograph face. If some signal is then detected from the specimen it can be used to brightness-modulate the display and so produce an image in which the contrast depends upon the variation of this signal as the probe scans across the specimen surface. The most commonly used form of scanning electron microscopy (SEM) employs the emitted true secondary electrons as the signal. Other forms use the current flowing from the specimen to ground (often called the absorbtion current), the electrons emitted with energies near to the elastic peak, or the emitted characteristic X-rays.

If an electron spectrometer is inserted between the specimen and the electron detector and the spectrometer is tuned to pass only electrons in a particular Auger peak, then the instrument becomes a scanning Auger microscope (SAM). Such a device is sketched in Fig. 2.12(b). Because the escape depths of Auger electrons are so small, SAM is a method which enables images to be made in which the contrast depends upon the variations from place to place across the surface of the amount of the chosen chemical element in the top few atomic layers of the sample.

Because the Auger cross-sections are quite small compared with those for elastic scattering or true secondary electron emission, the realization of a SAM with good quality images depends upon high currents in the incident beam and long scan times for each complete image. This is because if N Auger electrons are detected in each picture point, the root mean square fluctuations in N are just $N^{1/2}$ and the signal-to-noise ratio is thus $N^{1/2}$. If a signal-to-noise ratio of 10 is acceptable then 100 Auger electrons must be detected and with typical spectrometers ($\Omega = 1$ per cent of 2π sr) and Auger cross-sections this means that 10^8 electrons must be incident upon the sample. A crude image might contain 10^4 picture points obtained in, say, 100 s, so the primary beam current must be at least 10^{10} electrons s^{-1} or 1.6×10^{-9} A. These long picture acquisition times mean that it may be convenient to collect the numbers corresponding to the electrons counted at each picture point in the memory of a computer controlling the microscope. The brightness-modulated Auger image can then be displayed at the user's convenience after acquisition is finished. The spatial resolution obtainable depends upon the diameter of the region where the incident beam strikes the sample, and this diameter rises, because of source brightness and space charge effects, as more beam current is demanded. The outcome of the engineering compromises involved is that, for a spatial resolution below about 100 nm, field electron emission is the required source of electrons because of the high brightness attainable (see p. 118). For less demanding specifications, electron guns with thermionic sources are used.

The ability to spatially resolve Auger information has been very useful in many areas in technology. An example is given in Plate I for a layer structure composed of 10 nm of Co on 10 nm of Co_2Si on 10 nm of CoSi on a silicon substrate. A bevel has been cut into the surface of this structure with a computer controlled 2 keV Xe^+ beam. This leaves a free surface at an angle of about 1 mrad to the original surface. As a result each layer exposes a relatively wide region which can be observed using a variety of surface techniques which may not have a sufficiently small spatial resolution to measure the layers edge on. An image of the complete bevel is shown in Plate I(a), which was formed in a Auger electron microscope using a 20 keV beam of electrons scanning the surface and an energy analyser tuned to pass 100 eV electrons leaving the sample. The slope of the bevel is lowest at the right-hand side of the crater cut by the Xe^+ ions, and there are steeper sides at the top, bottom, and left-hand sides. Since the sample contains Co and Si, three Auger images are shown in Plates I(c), (d), and (e) which are formed using the peak heights of the Co

MVV peak at 45 eV, the Si LVV peak at 85 eV, and the Co LMM peak at 766 eV. The four layers in the sample can be discerned in the contrast in these Auger images. Plates I(e)–(h) show a low kinetic energy section of the electron spectrum from each of these four layers. (e) and (h) contain only the Auger peaks expected from pure Co and Si. The increasing size of the Si peak and the decreasing height of the Co peak is clear in these four spectra as the beam passes from the pure Co layer at the surface to the Si substrate. Quantitative analysis of the chemical composition of the two intermediate layers using the methods outlined below reveals that the layer intended to be Co_2Si is indeed near to that stoichiometry. The layer intended to be CoSi is, however, rather far from this composition, being rather nearer to an average composition of Co_xSi_{1-x}, with x near to 2, but with steadily changing Co concentration.

Quantitative analysis by AES and XPS

The quantitative estimation of elements at a surface is difficult with any of the electron spectroscopies in most circumstances, except when the element is uniformly distributed above or if the element is known to be uniformly distributed through the whole sample. Examples of practical situations are indicated in Fig. 2.13, and it is easily seen that measurement of Auger peak height alone will not resolve the difference between, for example, Fig. 2.13(d) and Fig. 2.13(e). Other techniques have to be brought to bear in addition to electron spectroscopy to help resolve the difficulties introduced by the existence of topography and spatial variations in the chemical distribution.

Nevertheless, it has been possible to develop techniques for the calculation of surface chemical composition for the simplest surface geometries in Fig. 2.13 when using AES or XPS. The methods are discussed fully in Briggs and Seah (1990). The most direct approach might be to measure the area under a peak in the Auger spectrum from each element in the surface and then to use eqn (2.6). This would be an absolute determination of the composition requiring knowledge of the ionization cross-sections ϕ, the escape depth τ of the Auger electrons, and the incident beam current i_0. Such information is not known accurately and may not even be available. A simpler approach is to measure the ratios of the areas or heights of corresponding Auger peaks in the spectra from the sample of unknown composition and an elemental standard or a standard of a compound with known surface composition. If these measurements are made in the same instrument with the same sample/electron gun/analyser geometry and the same beam current then many factors cancel out of eqn (2.6). This is now standard practice for quantitative analysis by electron spectroscopy. The mathematics of this approach are described in full in Briggs and Seah (1990), and the result for the atomic fraction X_A of element A in a multi-element sample is given by

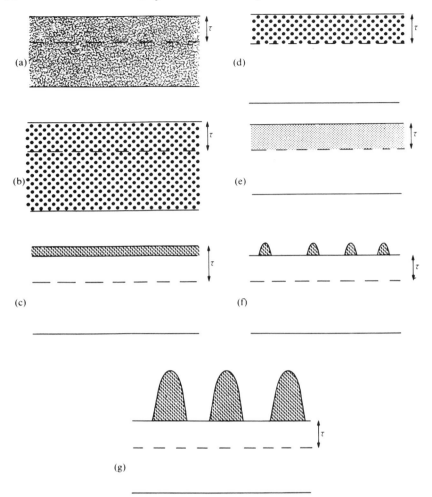

Fig. 2.13 Different arrangements of atoms of type A with substrates of type B which may occur in practice and which complicate the quantitative use of AES or XPS. (a) Uniform solution of atoms of A in matrix of B—e.g. an alloy. (b) Uniform distribution of clumps of atoms of A in matrix of B—precipitation. (c) Solutions of atoms of A only inside in the escape depth τ in B—surface segregation. (d) Clumps of atoms A inside the escape depth τ in B—surface precipitation. (e) Uniform thin layer of A upon B—a thin film. (f) Islands of atoms A upon B—e.g. some thin films in the initial stages of growth. (g) Tall needles (of height greater than τ) of A upon B—e.g. acicular growth of one substance upon another.

$$X_A = \frac{I_A/I_A^\infty}{\sum_B F_{BA}^A I_B/I_B^\infty}. \tag{2.9}$$

In this equation, I_A is the Auger peak area or height (the signal) from the sample and I_A^∞ is the corresponding signal from the elemental standard. The sum in the denominator is over the ratios of the signals from all the elements in the sample, and F is known as the *matrix correction factor*. This factor arises from the fact that the attenuation length of the Auger electrons, λ, the Auger backscattering factor r, and the density ρ of the sample depend upon its composition and so upon X_A. This means that eqn (2.9) has to be evaluated iteratively. The matrix correction factor for element A in the sample is given by

$$F_{BA}^A = \frac{r_M(E_A)r_B(E_B)\lambda_M(E_A)\lambda_B(E_B)a_A^3}{r_M(E_B)r_A(E_A)\lambda_M(E_B)\lambda_A(E_A)a_B^3}. \tag{2.10}$$

Here, the subscript M refers to a property of the sample and the subscripts A and B refer to the element A being determined and each of the other elements, which are labelled B in turn. The quantity a is the atomic size. As explained in Briggs and Seah, approximations for λ, ρ, and a can be derived for all elements and so eqn (2.8) can be evaluated iteratively using eqn (2.10). A computer program for this calculation of surface compositions from Auger spectra has been described by Walker *et al.* (1988).

Although these arguments have been presented for AES, a similar analytical approach works also for XPS, which is just as powerful for quantitative surface analysis. In either spectroscopy, accuracies in determining the surface composition in the range 5–10 per cent can be obtained routinely and with great care accuracies of the order of 1 per cent are possible. This kind of composition determination is important in studies of surface diffusion and segregation, as described later.

Secondary ion mass spectroscopy (SIMS)

In this technique the target surface is bombarded with a beam of primary ions having energies of several keV. Atoms and molecules are knocked (sputtered) out of the target and may emerge in their ground states, as excited particles, or as positive or negative ions. A fraction of these ions are extracted from the target region and passed to a mass spectrometer, where they can be separated according to their mass to charge ratio. Because sputtering probabilities can be high and mass spectrometers can be built with high sensitivities, this technique can give a much higher sensitivity to surface chemicals than Auger spectroscopy. However, it is a destructive technique because the surface atoms have to be knocked out of the solid in order to be detected and the sensitivity of the method varies from element to element and from matrix to matrix (for a given element) by several orders of magnitude.

The subject has been reviewed, for example, by Briggs and Seah (1992), who describes various ways of obtaining quantitative data using SIMS. Examples of mass spectra obtained from Mo metal are shown in Fig. 2.14 obtained by Benninghoven (1973) using 3 keV Ar^+ ions in the incident beam. Because this Mo surface had deliberately not been cleaned one can detect H^+ and H_2^+ from

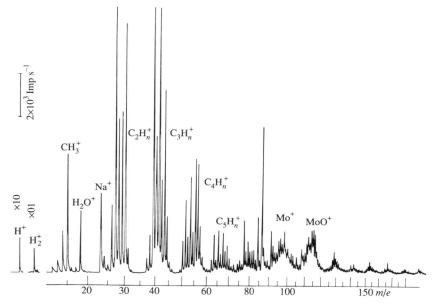

Fig. 2.14 Positive secondary ion spectrum of a molybdenum surface before any cleansing. Residual gas pressure 5×10^{-10} Torr. Incident beam current 10^{-9} A cm^{-2} of 3 keV Ar$^+$ ions. (By courtesy of A. Benninghoven.)

adsorbed hydrogen and hydrogen-bearing compounds, ions of $C_m H_n^+$ up to $m = 14$ from adsorbed hydrocarbons, and ions of the metal Mo and its oxide fraction MoO$^+$. If negative ions are detected the electronegative contaminants in the target can also be detected. By using a primary ion current as low as 10^{-11} A the erosion of the surface can be kept as low as 0.1 nm per hour. Another example of a SIMS spectrum, shown in Fig. 2.15, is from a high-temperature superconducting material—yttrium barium copper oxide. The range of peak intensities in this spectrum is typical of SIMS spectra at the time of writing—they span about five orders of magnitude. Few techniques have such a wide dynamic range, which is an advantage of SIMS additional to its high sensitivity. At the low end of the range the limits are set by the statistics of counting the number of ions with a selected mass. This can be improved by increasing the time of measurement, but this may be at the cost of eroding the sample so much that the incident ions begin to strike a region where the composition is different from the value it had at the start of bombardment. The upper limit is in principle set by 100 per cent of the mass number being detected. This range can span as many as seven or eight orders of magnitude in the concentration of the selected species. No other spectroscopy can approach this span.

The presence of molecular ions amongst the sputtered species makes this

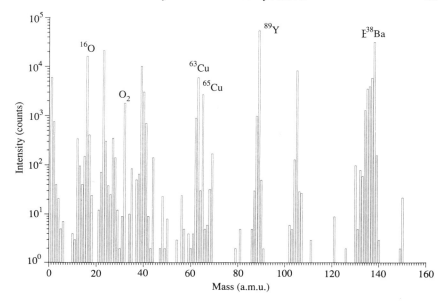

Fig. 2.15 SIMS spectrum of a high temperature superconductor $YBa_2Cu_3O_7$. EVA 2000 SIMS, primary ions 4 keV 0_2^+. (By courtesy of Dr M. Dowsett, University of Warwick.)

method particularly valuable in the study of molecular surfaces and molecular adsorbates upon simpler surfaces, because the SIMS spectrum has a characteristic form which can be related to the source of molecules by using standards.

Quantitative analysis by SIMS

The steps between measuring the size of the peaks in a SIMS spectrum and deriving the concentrations of the elements present are very tricky indeed. The simplest procedure is to compare the peak heights in the sample of unknown composition with the corresponding peaks from a standard sample of known surface composition. However, several factors conspire to make this less than straightforward. These include:

1. The surface and bulk compositions of the standard may not be identical. This can be the case because an ion beam used to clean the surface of the standard may sputter the atoms at different rates, depending upon their atomic numbers and the environments in which they are situated. This is called *differential sputtering*. Another possibility is that the surface and bulk compositions are different because *segregation* of one atomic species occurs to

or from the surface at a higher rate than any other species. This kind of difficulty can be minimized if a standard can be found which has a composition very close to that for the unknown sample. Then the standard and the unknown can be treated in the same way and remain comparable. An example of this kind of situation is the surface analysis by SIMS of semiconducting compounds in the III–V family. Here, a useful standard is a stoichiometric crystal of a binary compound such as GaAs. This can be cleaved *in situ* on a (110) crystal plane. This plane has an equal occupancy of Ga and As atoms. If the unknown has a composition only a few atomic per cent away from that of the stoichiometric compound, then quantitative analysis is possible with some confidence.

2. The transfer of momentum and energy from the incident ions to the atoms in the solid can modify multi-component solids in quite striking ways. Thus, for instance, a particular kind of atom that was originally on the surface can be driven into the solid and away from the exposed surface. Alternatively, the strains in the solid introduced by the incorporation of the bombarding species under the surface can promote diffusion of some kinds of atoms towards the surface!

3. The yield of ions from the surface which reach the mass spectrometer and are detected is very strongly dependent upon the environment of the ejected ion just before it left the solid. Incident or ejected ions may be neutralized before or after the ejection event—an effect which depends upon the electronic structure of the participants in the collision. The yield of a particular kind of emitted ion using a second kind of incident ion may vary by orders of magnitude, depending upon the composition of the solid from which the secondary ion is ejected and the local environment if that ion. For this reason it is usually referred to as a *matrix effect*.

Because of the sensitivity of SIMS and its wide range of applicability in surface science there is a worldwide research effort to try to model the secondary ion generation process. The model has to take into account the sputtering and mixing effects of the primary ions, the ionization probabilities of the atoms in the solid, the escape probabilities of the secondary ions, and the sensitivity of the detector. In practice, of course, the measurement is one of the height of a peak in a SIMS spectrum and the problem is to use a model to go backwards to deduce the composition of the solid from which this peak arose. It is a member of a general class of models in which *inversion* is required. It is computationally more rigorous to go forwards from a known solid to predict the heights of features in a spectrum than it is to go backwards in the way required. Some large computer programs are under development now to try to solve this problem. The subject is thoroughly reviewed in Briggs and Seah (1992).

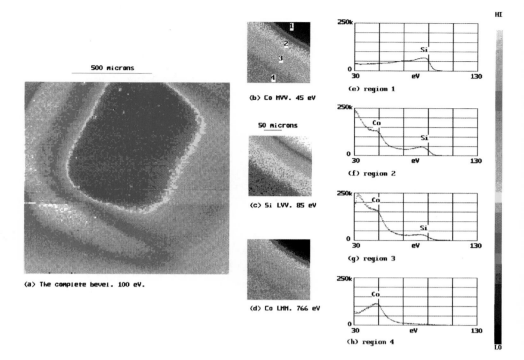

500 microns

50 microns

(a) The complete bevel. 100 eV.

(b) Co MVV. 45 eV

(c) Si LVV. 85 eV

(d) Co LMM. 766 eV

(e) region 1

(f) region 2

(g) region 3

(h) region 4

Plate I Energy-analysed and Auger images of a bevel cut with a computer controlled beam of xenon ions in a Co/Co_2 Si/CoSi/Si layer structure. (a) shows an image whose contrast is formed with 100 eV electrons leaving the sample which was scanned with a 20 keV beam of electrons and a beam current of about 6 nA. (b), (c), and (d) show the Auger images formed at the energies indicated. Because the Co MVV and the Si LVV peaks are superimposed on a rapidly changing background of secondary electrons, both (b) and (c) are calculated from three images. One image is collected on the Auger peak and two separate images are obtained at two energies on the background in the electron spectrum just above the Auger peak. The peak height is then estimated by extrapolating the background at each point in the image to the Auger energy. The Auger image is then the difference between the image obtained on the peak and that extrapolated under the peak from the two background images. The spectral background at the Co LMM energy is sufficiently flat that the Auger image can be obtained by subtracting the image measured on the background just above the peak from that measured on the peak. The four spectra (e) to (h) are low kinetic energy regions of the full spectrum obtained from regions within each of the four layers in the structure. The spectral background has been reduced by subtracting from the region shown here a function of the form AE^{-m} determined by non-linear least squares fitting to a featureless region of the spectrum between 105 and 160 eV. The progression in the sizes of the Si and Co peaks is clear through the layer thickness. Close examination of the energies of the Si LVV peaks in the mixed layers reveals that there is a shift of 2.5–3 eV from the position of the bulk Si LVV peak. This indicates that chemical bonding has occurred between the Co and the Si, which has modified the densities of occupied electron states in the Si.

(a)

(b)

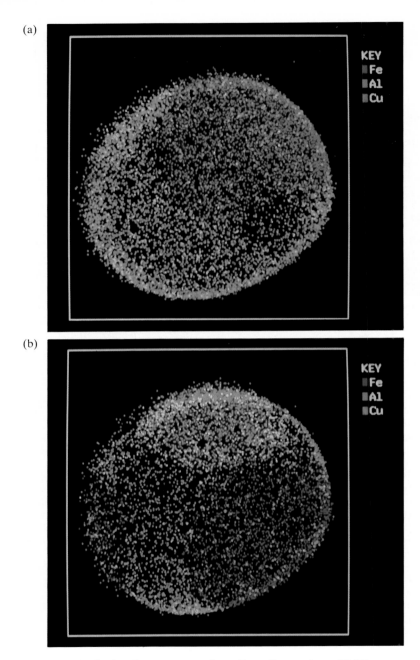

Plate II Element distributions from two areas about 15 nm diameter in an AlNiCo$_2$ permanent magnet alloy. (a) A region which is rich in Co and Fe in the top right and rich in Al, Ni, and Cu in the bottom left. (b) A region containing a 5 nm particle of a Cu rich phase. (By courtesy of Dr A. Cerezo, University of Oxford.)

SIMS, Auger, and XPS microscopies compared

Several of the spectroscopies described above can be adapted to become microscopies. This is particularly important if the surface of interest is likely to be inhomogeneous—i.e. it varies in composition or crystallography from place to place. Even if the surface is believed to be homogeneous it can be salutory to use a microscope to examine it. Belief is one thing and proof is another!

There are two kinds of technique for microscopy. The most common in surface work is a scanning beam method, in which the incident particles or radiation are focused on the surface and then scanned across it in a discrete, digital fashion or a continuous, analogue fashion. At each point on the surface some emitted particle or radiation is detected. The signal so detected can be used to modulate the brightness of a cathode ray tube display which is being scanned by an electron beam moving in synchronism with the beam striking the sample. This is the principle of the scanning electron microscope (and the scanning Auger electron microscope) described above. Rather than modulate a synchonously rastered display tube, it is convenient and useful to scan the sample surface in a digital fashion and store a number proportional to the intensity of emitted radiation or the number of emitted particles. Each entry is commonly referred to as a *pixel*. This is particularly powerful because the computer representation of the image can be manipulated or processed with a wide variety of algorithms to enhance the contrast, extract data along a line of interest on the surface, or compare one image with another. The methods of image processing developed for interpreting and displaying satellite and space vehicle pictures are of considerable value here (see, for instance, Niblack 1986).

A second technique is analogous to optical microscopy in that a lens system is used to focus an image of some illuminated area of the sample on to a detector. The whole image may then be photographed or transferred electronically to a display or to the memory of a digital computer. This has the advantage that the data about the sample are collected simultaneously from all of the area being illuminated. By contrast, the scanning technique collects data about the sample sequentially point by point. Which technique is the most appropriate depends upon the extent to which the incident beam damages the surface which is being studied, what spatial resolution is required, and whether or not imaging systems are available for the radiation or particles leaving the surface.

Electron beams focused on a sample are particularly easy to scan, but energy-selective techniques such as are provided by the spectrometers described earlier in this chapter tend to suffer from very large spherical aberration effects. This means that scanning is the preferred technology because the spatial resolution obtainable will then be determined by the diameter of the focused beam falling on the sample. Ion beams can also be focused and scanned quite conveniently and so the SIMS methodology can also be adapted into a scanning microscopy.

The modification of XPS into a microscopy is less straightforward. The area of the sample illuminated by X-rays in a conventional XPS system is quite large (several square mm) and so a scanning method would have very poor spatial resolution. The aternatives are:

(1) to use an X-ray monochromator or curved mirror to focus the X-rays on to the sample and to move either the sample itself or the X-ray beam across the surface in a raster pattern; or

(2) to accept the illumination of a large area of the sample and view the photoelectrons with a spectrometer tuned to the energy of interest but capable of imaging the emitting region onto a detector surface.

Both of these methods have been used, but resolutions of the order of a few μm have been the best possible. In case (1), it is difficult to focus adequate X-ray intensity into a smaller spot on the surface. In case (2), the spherical aberrations in the electron lenses limit the size of the smallest detail that can be resolved.

SIMS adapted to a microscopy is very attractive because the high sensitivity of SIMS to small amounts of material means, for example, that the spatial distributions of impurities in semiconductors can be explored. Indeed, by using the mass-resolving capability of SIMS, together with the scanning of the incident ion beam and the fact that a surface can be progressively eroded with the ion beam, as described below, a three-dimensional reconstruction of the distribution of chemical elements in a surface can be derived. There are compromises between the sensitivity of the microscopy to a particular impurity and the spatial resolution obtainable, because the smaller the diameter of the incident ion beam the better the spatial resolution but the smaller the number of secondary ions ejected for measurement. Further, the difficulties of quantification of SIMS outlined above apply equally to the imaging situation where quantitative interpretation of the amount of each impurity at each place in the surface is required.

SAM is more readily quantified using the kind of expressions described in the discussion of quantitative AES above. However, the sensitivity of AES is not usually as high as that of SIMS. Because SIMS is inherently a consumer of the material of the sample and AES is not, then as the incident beam becomes smaller the sensitivity of SIMS reduces but the sensitivity of AES stays roughly constant. Therefore, SAM becomes more sensitive than scanning SIMS below some critical incident beam size. This is usually about 100 nm.

Discussion here has been confined to microscopies which are chemically specific. In Chapter 3 a non-scanning diffraction technique is described which can be used to image the lateral variations in the crystallography of a surface with very high spatial resolution.

Depth profiling

As mentioned above, the bombardment of a surface with ions of a noble gas such as argon can result in the slow but progressive removal of atoms. If photoelectron or Auger spectroscopy or SIMS of the surface is carried out as this etching of the surface proceeds, then the variation of the chemical composition of the solid with depth from the original surface can be derived. This procedure is often called depth profiling or ion-beam etching.

This is extremely useful as a method for examining the chemical distribution near surfaces or interfaces and has been extensively developed both for the analysis of semiconductor devices and corrosion processes. An example of such an analysis is shown in Fig. 2.16, where the distribution of germanium through a superlattice between layers of pure Si and layers of $Si_{0.853}Ge_{0.147}$ can be seen to have a depth resolution of less than about 5 nm. This is an extremely good depth resolution. The difficulty associated with this technique is that the different atomic types in a heterogeneous material may be removed at different rates by the ion bombardment. Thus, the depth profile is modified from the original chemical distribution, owing to the inhomogeneous removal of material and the topographical effects which develop. Nevertheless, depth resolutions of a few nm can be achieved.

The problems of quantitation mentioned above in the discussion of SIMS are just the same here. Now, however, the inversion problem has to be solved at each depth in the solid. It is easy to see that efficient and fast computational algorithms are essential if a depth profile is to be converted from a plot of secondary ion counts versus time of bombardment to what is really required— a plot of composition versus depth.

In spite of the reservations expressed here about the difficulties that arise in trying to obtain quantitative compositional information using SIMS, it should be stressed that the sensitivity of the technique is sufficiently high compared to most other methods that it is well worth collecting and trying to interpret this kind of information. If there is little alternative, then some information is always better than no information—even imperfect information can lead to new ideas which can be tested in different ways!

The atom probe

By combining a field-ion microscope (see Chapter 3) with an aperture in its imaging screen followed by a time of flight mass spectrometer, Muller and Tsong (1969) were able to detect and identify (by the ratio m/e) individual atoms. This has the ultimate chemical sensitivity, but it is limited to rather special experimental geometries suitable for the field ionization of atoms. Cerezo *et al.* (1989) have developed these ideas to make an instrument they call a POSAP, or *position sensitive atom probe*. The instrument is shown schematically in Fig. 2.17. Successive layers of atoms are stripped from the

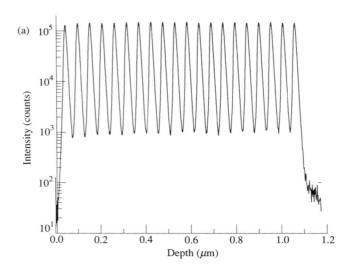

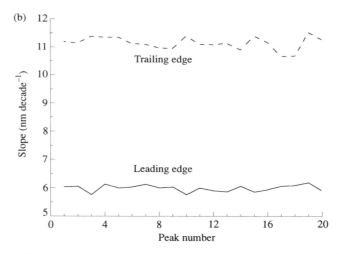

Fig. 2.16 (a) SIMS depth composition profile through a sample containing alternate layers of Si and $Si_{0.853}Ge_{0.147}$. The sample contains 20 periods of layers of 44.6 nm thick Si followed by 8.6 nm thick SiGe. It is a strained layer superlattice. The depth profile was generated by bombarding the sample with a 4 keV beam of O^+ ions in normal incidence. (b) shows the slopes of the leading and trailing edges of each peak in (a). The difference in the leading and trailing slopes is due to the SIMS process changing the sample as it is measured, and is not a property of the sample as it was fabricated. (By courtesy of Dr R. D. Barlow and Dr M. Dowsett, University of Warwick, for the SIMS data, and Dr R. A. Kubiak and Prof. E. H. C. Parker, University of Warwick, for the MBE growth of the layer structure.)

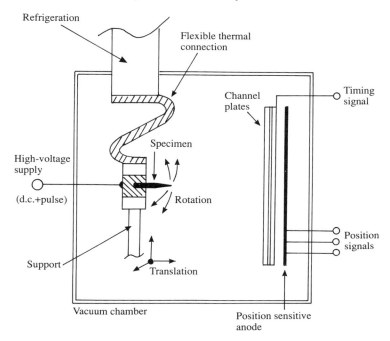

Fig. 2.17 A diagram of the principles of the position sensitive atom probe (POSAP) (after Cerezo *et al.* 1989). (By courtesy of Dr A. Cerezo, University of Oxford.)

sharp tip of the sample and the time is measured for each kind of atom to reach the position sensitive detector behind a pair of micro-channel plates. Since the time of flight is determined by the mass to charge ratio of the atoms leaving the sample, the technique identifies each atom arriving at the position sensitive detector. Further, the position of each pulse detected depends upon the position on the sample whence the atom originated. Areas as large as 10–20 nm diameter on the sample surface can be identified with atomic spatial resolution.

An example of the kind of results obtainable from this instrument is shown in Plate II. Here, element distributions from two different areas on the surface of an AlNiCo permanent magnet alloy are shown. The inset beside each micrograph shows the area analysis of the amount of each element in the imaged area. This can be measured by counting the atoms of each kind!

A case study—NiCrAl + O

An interesting case study of the application of electron spectroscopies to a practical problem in materials science is provided by the example of layers of protective oxide on the surface of superalloys. These metals are ternary or

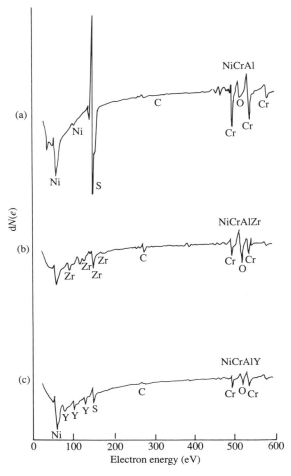

Fig. 2.18 Differentiated Auger spectra of NiCrAl, NiCrAlY, and NiCrAlZr alloys after annealing at 750°C in UHV. (After Smialek and Browning 1986).

quaternary alloys of the elements Ni, Fe, or Co combined with Al and Cr. Often small amounts of reactive elements like Zr, Yt, or Hf are added because it has been discovered empirically that these elements cause the surface oxide on the alloy to adhere more firmly. These materials are very important for high-temperature components operating in oxidizing atmospheres. Thus, they are commonly used for the blades of the turbines in jet engines. In this application the blades may be operated at about 1100°C in air at atmospheric pressure. The life of the blades is prolonged by the growth of a dense layer of oxide which forms when they are first heated in air. Subsequent heating still causes oxidation, but at a much slower rate because of the protective effect of the initial oxide, which prevents oxygen-bearing molecules from reaching unoxidized metal. However, the oxide does continue to grow during the life of the blades until it becomes so thick and is so stressed that it falls off the alloy and fresh metal is exposed to air. This *spalling* process causes gradual erosion of the alloy until it breaks. If an alloy composition can be found which prolongs the life of the alloy before blades have to be replaced then this has clearly been a useful contribution to both the costs and the safety of flying!

Before surface analytical techniques were applied to the study of oxide adhesion and the identification of the role of the reactive additives there were several theories for these effects. These included:

1. The coefficients of thermal expansion of the oxide coating and the alloy can be very different. A large stress will build up in the oxide due to this difference as the material cools to room temperature from the high temperature of operation of the blades. This might cause the oxide to break away from the metal so exposing fresh, unoxidized regions. It can be postulated that addition of the reactive metal to the alloy may lead to the formation of an interfacial layer between the oxide and the alloy. If this layer has a coefficient of thermal expansion intermediate between those of the oxide and the alloy then it may allow a lower gradient of thermal strain across the interface and so a reduced tendency for the oxide to break off.

2. The addition of the reactive element may result in a change of the morphology of the interface. One possibility is that '*tongues*' or '*pegs*' of oxide may form, penetrating into the alloy. This two-dimensional comb-like structure may give more strength to the bonding between the oxide and the alloy.

3. The oxide layer consists principally of Al_2O_3. The oxide may grow at an elevated temperature by diffusion of the aluminium out of the alloy and along dislocations in the oxide film together with inward diffusion of oxygen from the atmosphere into the oxide along grain boundaries. Perhaps the reactive element additive to the alloy blocks the outward diffusion of aluminium by migrating to the dislocations in the oxide.

One interesting study which may help to distinguish between these possibilities is to try to identify the location of the reactive metal additive in the oxidized system. Is it in the scale, in the alloy, at the interface or in some combination of these positions? Is it homogeneously or inhomogeneously distributed in particular regions? AES and/or XPS, combined with depth profiling techniques, might reasonably be expected to be helpful in resolving these questions. Work of this kind has been carried out by a number of authors and the results have been reviewed by Smialek and Browning (1986). The Auger spectra of NiCrAl alloys doped with 0.5 per cent (by weight) of Zr or Y are compared with those observed from an alloy containing no reactive element dopant in Fig. 2.18. In these data all three samples had been annealed at 750°C for 20 min in UHV. A large sulphur peak is evident in the Auger spectrum of the undoped sample in spite of the fact that bulk analysis had shown that it contained less than 10 p.p.m. of sulphur. Just as is found with many nickel-based alloys, sulphur segregates to the surface. However, both the Y and the Zr additives result in alloys which show very much smaller sulphur Auger peaks, together with noticeable Y or Zr peaks.

The interpretation of these results is that the sulphur segregates to the surface of the undoped alloys and results in a weaker bond to any oxide layer than would have been formed with a clean alloy surface. When the reactive element is added to the alloy the diffusion of sulphur to the surface is inhibited, presumably by the formation of larger molecules of sulphur bound to Y or Zr. With less sulphur at the oxide-doped alloy interface the bond to the oxide is stronger and the adhesion of the oxide is enhanced. It is not untypical of surface experiments that they lead to conclusions at variance with all the theories extant before the experiments were carried out! Happily, however, the new results often throw a revealing light on the situation, which leads to new, and sometimes more satisfactory, theories being created. This often is one of the satisfactory aspects of the advance of materials science.

3
Surface structure

Having determined what types of atoms are present upon a surface, the next important problem is to discover their arrangements with respect to each other and with respect to the underlying atoms of the solid. There are two parts to this problem—the determination of the symmetry of the surface atomic arrangement and the determination of details of the atomic positions. In bulk investigations, the former is normally carried out by using simple observations of the diffraction pattern obtained when a beam of X-rays, neutrons, or electrons is scattered from a single crystal of the sample. These observations yield information about the symmetry of the repeating unit of the structure, the unit cell, and the size and shape of this cell. The latter involves measurements of the intensity of diffracted beams and the comparison of these intensities with those predicted from postulated models of the structure. If successful, the result is a complete description of where each kind of atom is situated within the unit cell.

Bulk techniques for structure analysis

X-ray diffraction is by far the most common method for studying bulk structures. X-rays are scattered by the charge distribution in and around atoms, and because this scattering is very weak they can penetrate materials very deeply and the bulk structure can be probed. By the same token, the small atomic scattering cross-sections of atoms for X-rays result in a relative insensitivity to surface atoms. Using high-intensity laboratory X-ray sources and sensitive X-ray detectors, it has been possible to study the structure of metal films 10 nm thick, but the difficulties of extending this work to smaller thicknesses are considerable. Further improvements to the surface sensitivity of X-ray scattering can be obtained by using grazing incidence of an intense beam (see later). In spite of these reservations about the use of X-rays for surface structural studies, it helps to understand the procedures of structure analysis if the principal steps in a bulk X-ray structure determination of a simple solid are examined in a simplified way.

A possible sequence of events for a structure determination is outlined in Fig. 3.1. Having obtained a single crystal of the sample and determined its composition (a combination that may be by no means trivial) a Laue diffraction pattern is obtained (e.g. Wormald 1973). Combining these observations with the Ewald sphere construction (Fig. 3.2), the unit cell of the structure can be determined. In finding the unit cell it is particularly useful to be aware of systematic absences. These arise whenever the direction of

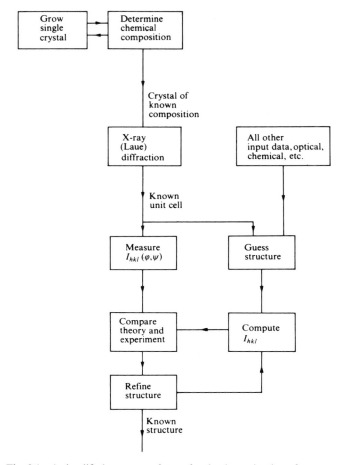

Fig. 3.1 A simplified sequence of steps for the determination of a structure.

diffraction corresponds to destructive interference between the scattered waves. These absences can be found by using the *structure factor* F_{hkl} of the unit cell. The intensity I_{hkl} of a spot with Miller indices (hkl) is related to F_{hkl} through

$$I_{hkl} = |F_{hkl}|^2, \tag{3.1}$$

and, for cubic structures, F_{hkl} is given by

$$F_{hkl} = \Sigma_p f_p \exp 2\pi i(\mu h + \nu k + \omega l). \tag{3.2}$$

In this equation the sum is over the atoms in the unit cell; f_p are the atomic scattering factors for each type of atom p in the structure and μ, ν, and ω are the positions of the atoms in the until cell expressed as fractions of the cell sides.

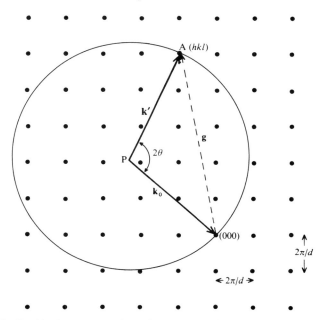

Fig. 3.2 The Ewald sphere construction. The wave vector k_0, of the incident radiation is drawn on a scale diagram of the reciprocal lattice of the structure. The tip of k_0 is at the origin (000) of reciprocal space. A sphere drawn with radius k_0 and with centre P is the Ewald sphere. If a point on the reciprocal lattice lies on the surface of the Ewald sphere, the condition for elastic scattering (the Bragg equation) is satisfied and diffraction will occur with the scattered beam having wave vector k'. For monochromatic X-rays and arbitrary k_0 it is clear that very few diffracted beams are excited (only beam PA above). The case shown is for a simple cubic lattice. 2θ is the scattering angle. g is the reciprocal lattice vector and (hkl) the Miller indices of the point on the Ewald sphere. By using $|k| = 1/\lambda$, $|k_0| = |k'|$, and by dropping a perpendicular from P to g it can be shown that the construction is equivalent to

$$n\lambda = 2d \sin \theta \qquad \text{(Bragg's law)}$$

or

$$k' = k + g,$$

where λ is the wavelength and $n = (h^2 + k^2 + l^2)^{1/2}$.

Some examples of simple systematic absences are:

(1) in the f.c.c. system, where there are no diffracted beams with (hkl) mixed even and odd;

(2) in the body-centred cubic (b.c.c.) system only beams with ($h + k + l$) even are found—all others are systematic absences;

(3) in the sodium chloride-type structure there are bright beams when (hkl) are all even, dim beams for (hkl) all odd, and systematic absences for all other (hkl).

These results can be demonstrated using eqn (3.2), and similar rules exist for all structures. It is possible to learn to recognize these features of diffraction patterns with practice.

By measuring the angles between diffracted beams and knowing the wavelength λ, the sides of the unit cell can be determined by using Bragg's law.

The difficult step in structure analysis comes after the identification and measurement of the unit cell, when it is decided to try to find out where the atoms are within the unit cell. Figure 3.1 indicates the kind of procedure that can be adopted to tackle this part of the problem. The difficulty arises from the fact that the experimental measurement must be of the intensity of the diffracted wave and this involves losing information about its phase. Because of this loss of information, the diffraction pattern cannot be unravelled to reveal the structure directly, but other information, obtained in different experiments, has to be employed. For instance, chemical reactivity and spectroscopic observations can give information about bonding between the atoms present, and the natural occurrence of certain crystal faces suggests that these have low free energies and this observation may imply that particular combinations of atoms may be in these faces. All such guides are used together with the unit cell determination to propose a model structure and make predictions, through equations like (3.2), of the intensities of diffracted beams as a function of the diffraction geometry (Fig. 3.3). These predictions are compared with a detailed set of measurements of the intensities I_{hkl} of a large set of beams (hkl) as functions of the angles ϕ and ψ explained in Fig. 3.3. This comparison enables more refined models of the structure to be made, and the loop is repeated until satisfactory agreement between theory and experiment is obtained.

The same procedure can be used to solve structures by observing the elastic scattering of neutrons in crystals. However, the scattering cross-sections of atoms for neutrons are lower than those for X-rays and, consequently, neutron scattering is even less surface-sensitive than X-ray scattering.

A form of incident wave which is much more strongly scattered by atoms then either neutrons or X-rays is one of electrons. The atomic scattering probabilities for three processes are indicated as a function of their energies in Fig. 3.4. Electrons are usually at least 10^3 times more strongly scattered than X-rays and are particularly strongly scattered if their energies are below 1000 eV. It is this strong scattering cross-section that makes electrons suitable probes for surface structure studies.

Surface methods using electrons

The two important techniques for surface structure analysis using electrons are low-energy electron diffraction (LEED) and reflection high-energy electron diffraction (RHEED). Both methods can be used to determine the

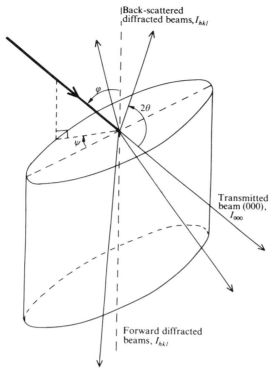

Fig. 3.3 Diffraction geometry from a cylindrical sample. ϕ is the angle of incidence measured from the normal to the surface upon which the radiation impinges. ψ is the azimuthal angle. It is measured from a reference direction in the surface (perhaps a low-index crystal direction) to the projection of the incident beam on to the surface. 2θ is the scattering angle.

periodic two-dimensional arrangement of atoms at the surface—the unit mesh. A number of such arrangements are possible and it is useful to have some notation to describe this unit mesh before discussing the experimental methods.

Notation for surface structures

Just as bulk, triperiodic structures can be divided into 14 groups corresponding to their different Bravais lattices (Rosenberg 1974), so two-dimensional periodic structures can be grouped into five types of surface net. These five nets describe all possible diperiodic surface structures. It is convenient to specify the surface itself by the Miller indices (hkl) of its normal. Then Fig. 3.5 shows the unit areas or *unit meshes* of the five possible nets. The complete net can be generated by translating the unit mesh parallel to the vectors $\mathbf{a}_{1s}$ and $\mathbf{a}_{2s}$ an

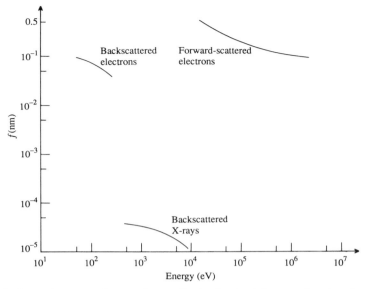

Fig. 3.4 Approximate atomic scattering amplitudes for aluminium as a function of energy of electrons and X-rays. As the scattering amplitude increases, so the mean depth sampled by the electrons decreases.

integral number of times. The subscript s will be used for a surface mesh and the subscript b for a bulk exposed plane.

It is convenient to be able to relate the translations $\mathbf{a}_{1s}$ and $\mathbf{a}_{2s}$ of a surface net area A to the translations $\mathbf{a}_{1b}$ and $\mathbf{a}_{2b}$ of a second net of area B. This is required, for instance, if a reconstructed surface is to be described in terms of the bulk exposed plane (Chapter 1) or of a monolayer of some surface deposit has a different unit mesh from that of the solid surface upon which it is adsorbed. This relationship is made most powerfully by using a transformation matrix $\mathbf{M}$ such that

$$\mathbf{a}_s = \mathbf{M}\mathbf{a}_b, \tag{3.3}$$

where

$$\mathbf{M} = \begin{bmatrix} m_{11} & m_{21} \\ m_{21} & m_{22} \end{bmatrix} \tag{3.4}$$

and thus

$$\mathbf{a}_{1s} = m_{11}\mathbf{a}_{1b} + m_{12}\mathbf{a}_{2b}$$

$$\mathbf{a}_{2s} = m_{21}\,\mathbf{a}_{1b} + m_{22}\mathbf{a}_{2b}. \tag{3.5}$$

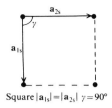

Square $|\mathbf{a}_{1s}|=|\mathbf{a}_{2s}|$ $\gamma=90°$

Rectangular   Centred rectangular

$|\mathbf{a}_{1s}| \neq |\mathbf{a}_{2s}|$ $\gamma=90°$

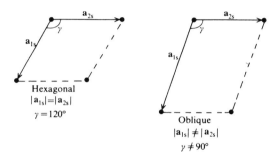

Hexagonal
$|\mathbf{a}_{1s}|=|\mathbf{a}_{2s}|$
$\gamma=120°$

Oblique
$|\mathbf{a}_{1s}| \neq |\mathbf{a}_{2s}|$
$\gamma \neq 90°$

Fig. 3.5 Unit meshes of the five possible surface nets.

Since the areas of the two unit meshes are given by

$$A=|\mathbf{a}_{1s} \times \mathbf{a}_{2s}| \tag{3.6}$$

and

$$B=|\mathbf{a}_{1b} \times \mathbf{a}_{2b}|, \tag{3.7}$$

eqn (3.5) can be inserted into eqn (3.6) to yield (using eqn (3.7))

$$A=B \det \mathbf{M}. \tag{3.8}$$

The values of det $\mathbf{M}$ can be used to define the type of superposition which exists between the surface mesh $\mathbf{a}_s$ and the bulk mesh $\mathbf{a}_b$. This definition and the cases of simple, coincidence, and incoherent superpositions are described in Fig. 3.6. Although an example is not shown in Fig. 3.6 for an incoherent superposition, cases of this type are found in practice and they can correspond to some cases of the phenomenon of epitaxial growth described in Chapter 6.

Although the matrix notation is powerful and general, many observed superpositions have been described by a simple notation devised by Wood (1964). Here the meshes $\mathbf{a}_s$ for the surface and $\mathbf{a}_b$ for the bulk are related by the ratios of the lengths of the translation vectors and by a rotation R expressed in degrees. Thus the meshes are related by an expression of the form $(\mathbf{a}_{1s}/\mathbf{a}_{1b} \times \mathbf{a}_{2s}/\mathbf{a}_{2b})R$. Two examples of this notation are illustrated in Fig. 3.7. If the deposit mesh is not rotated with respect to the substrate (or bulk) mesh then the rotation is simply dropped from the notation. Similarly, a statement as to whether the mesh is primitive or centred (Fig. 3.5) should be included in the notation, but it is usual practice to drop any symbol for a primitive mesh and to insert a lower case c before the statement of translation vector ratios for a centred mesh. This notation becomes clearer as examples are studied.

Diffraction from diperiodic structures

Because the periodicity along the surface normal is lost in a two-dimensional arrangement of atoms, the constructive interference of scattered waves cannot occur in this direction. This relaxation of the conditions for diffraction leads to the possibility of diffracted beams occurring at all energies, and hence the fact that a diffraction pattern can be observed at all energies and in any geometry. This can be understood by using again the Ewald sphere construction in a reciprocal lattice diagram for a two-dimensional net of atoms. A rather loose view of this reciprocal space diagram can be obtained by realizing that distances in reciprocal space are inversely proportional to distances in real space, and so if a triperiodic lattice is extended along one of its axes then reciprocal lattice points move closer togther along this axis. In the extreme of this extension only one plane of atoms is left (the others being removed to infinity) and the reciprocal lattice has become a set of infinitely long rods normal to the plane of atoms.†

The spacings $\mathbf{a}_1^*$ and $\mathbf{a}_2^*$ of the unit mesh of the reciprocal lattice are related to the spacings $\mathbf{a}_1$ and $\mathbf{a}_2$ of the real lattice by the equation

$$\mathbf{a}_i^* \cdot \mathbf{a}_j^* = 2\pi\delta_{ij}$$
$$\delta_{ij} = 1, \, i = j$$
$$\delta_{ij} = 0, \, i \neq j. \tag{3.9}$$

2π is chosen on the right-hand side of eqn (3.9) (instead of unity, as conventionally used in X-ray diffraction) because it enables the wave vectors $\mathbf{k}$

---

†A more rigorous method of deriving this result is via the use of the fact that the reciprocal lattice is the Fourier transform of the real lattice (Woolfson 1971). As the Fourier transform of a plane is an infinite rod normal to the plane, so the Fourier transform of a coplanar set of points is a set of infinite rods normal to the plane.

(a) SIMPLE det **M** integer

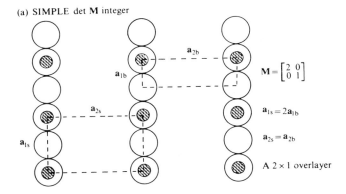

$$\mathbf{M} = \begin{bmatrix} 2 & 0 \\ 0 & 1 \end{bmatrix}$$

$\mathbf{a}_{1s} = 2\mathbf{a}_{1b}$

$\mathbf{a}_{2s} = \mathbf{a}_{2b}$

A 2 × 1 overlayer

(b) COINCIDENCE. det **M** a rational fraction

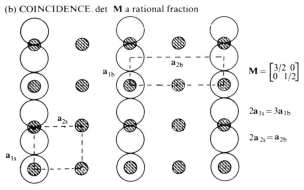

$$\mathbf{M} = \begin{bmatrix} 3/2 & 0 \\ 0 & 1/2 \end{bmatrix}$$

$2\mathbf{a}_{1s} = 3\mathbf{a}_{1b}$

$2\mathbf{a}_{2s} = \mathbf{a}_{2b}$

Fig. 3.6 Relationships between surface and bulk meshes. The simple and coincidence meshes are illustrated by the cases of deposit atoms (hatched circles) on the bulk exposed (110) plane of an f.c.c. material (open circles).

to be drawn upon the reciprocal lattice diagram and the Bragg equation (Fig. 3.2) to be employed directly. This is because the energy E of the incident electron beam is given by

$$E = (\hbar^2/2m)k^2 \tag{3.10}$$

and

$$k = 2\pi/\lambda. \tag{3.11}$$

The Ewald sphere construction can then be applied to the diperiodic diffraction problem, as shown in Fig. 3.8. Because of the loss of periodicity in one dimension, the reciprocal lattice rods can be labelled with only two Miller indices, h and k, and a general reciprocal lattice vector $\mathbf{g}_{hk}$ lies in the plane of the surface and is given by

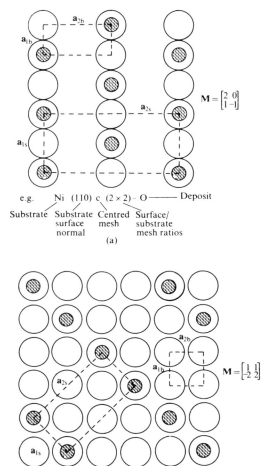

Fig. 3.7 Two notional examples of Wood's notation for surface structure compared with the matrix notation. In example (a) for a Ni(110) face exposed to oxygen the notation can be shortened slightly to Ni(110)c2-O because the deposit mesh is rectangular.

$$\mathbf{g}_{hk} = h\mathbf{a}_1^* + k\mathbf{a}_2^*. \tag{3.12}$$

Diffraction occurs everywhere the Ewald sphere cuts a reciprocal lattice rod and the diffracted beam can be labelled with the Miller indices (hk) of the rod causing it. Again, because of the loss of periodicity in one dimension, the vector diffraction equation becomes

$$\mathbf{k}'_{\parallel} = \mathbf{k}_{0\parallel} + \mathbf{g}_{hk} \tag{3.13a}$$

$$|\mathbf{k}'| = |\mathbf{k}_0|, \tag{3.13b}$$

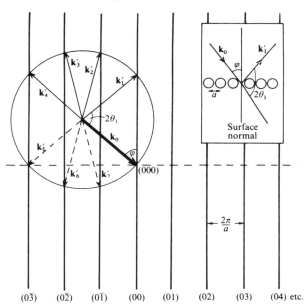

Fig. 3.8 The Ewald sphere construction (e.g. Wormald 1973) applied to diffraction from a square net of atoms of side a. In the case drawn, seven elastically scattered diffracted beams can be generated by incident wave vector $\mathbf{k}_0$ arriving at an angle of incidence ϕ to the surface normal. Four beams are shown backscattered from the surface; three beams are shown transmitted into the solid. More than this number of beams will occur because only that part of reciprocal space in the plane of the paper can be drawn. *Inset*: Real space diagram of the specular beam $\mathbf{k}_1'$.

where the subscript ∥ means the component of a vector parallel to the surface. Equation (3.13a) amounts to conservation of the component of momentum parallel to the surface, and its validity can be seen in Fig. 3.8. Equation (3.13b) amounts to the conservation of energy, used here because it is an elastic scattering process under consideration.

Reflection high-energy electron diffraction (RHEED)

If a beam of high-energy electrons is incident upon a flat surface in grazing incidence (Fig. 3.9) the diffraction pattern formed will be characteristic of the surface atomic arrangement because the component of the incident electron momentum normal to the surface is very small and thus the penetration of the electron beam will be small.

At high energies, the wavelength of the electron is small and the radius of the Ewald sphere is large compared to typical reciprocal lattice vectors. Thus, at 100 keV, $\lambda = 0.0037$ nm and $|\mathbf{k}_0| = 1700$ nm^{-1}, whereas $2\pi/a$ might typically be 20 nm^{-1}. Compared to the reciprocal lattice, the Ewald sphere is very large

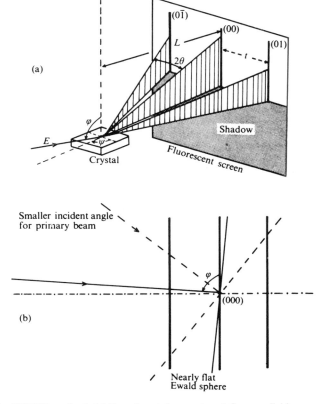

Fig. 3.9 The RHEED method. (a) Experimental geometry. A fine parallel beam of electrons is incident near $\phi = 90°$ upon a flat single-crystal surface. (b) Ewald sphere construction for RHEED.

and will cut the (00) rod almost along its length, as indicated in Fig. 3.9(b). Only those rods in the plane normal to the paper and containing the (00) rod will contribute beams to the diffraction pattern. This RHEED pattern will therefore consist of long streaks normal to the shadow edge of the sample and spaced by a distance t. If the separation between the fluorescent screen and the sample of some cubic crystal is L (the camera length) and the distance between streaks on the screen is t then

$$t = L \tan 2\theta$$

by geometry and

$$\lambda = 2d \sin \theta = \frac{2a}{(h^2 + k^2)^{1/2}} \sin \theta,$$

by Bragg's law for a sample with a square lattice of side a. As $\lambda \ll a$ in RHEED, then the values of 2θ are small and these equations can be simplified to yield

$$a = (h^2 + k^2)^{1/2} \lambda L/t \qquad (3.14)$$

All the parameters on the right-hand side of eqn (3.14) can be deduced or measured, and so a can be determined. The accuracy with which this can be done is determined by the accuracy with which t can be measured. This is determined largely by the length L. High-accuracy RHEED experiments are usually constructed so as to have long camera lengths for this reason.

Because the Ewald sphere is so large in a RHEED experiment it is necessary to change the diffraction geometry in order to find the arrangement of reciprocal lattice rods in three dimensions, and thus to define the unit mesh. As can be seen using Fig. 3.9(b), all the reciprocal lattice rods can be explored either by changing the angle ϕ by rocking the sample about an axis in its surface or by rotating the sample about its surface normal. The latter is to be preferred if constant surface sensitivity is to be maintained, as the rocking experiment changes the component of the incident electron momentum normal to the surface. If the azimuthal angle (ψ in Fig. 3.3) is varied, simple streak diffraction patterns can be observed when the incident beam is along directions of high crystal symmetry. An example of RHEED patterns from a Si(111) surface in two azimuths is shown in Fig. 3.10.

The surface sensitivity of the RHEED method is affected by two factors in addition to those described above. The simplest of these two is the surface roughness. If small bumps or needles of material stick out of the surface then the incident RHEED beam will pass through them. In doing so it will be diffracted by the three-dimensional atomic arrangement within the bumps. The diffraction process is then much more like diffraction in the bulk material and spots are observed instead of streaks. On the one hand, this phenomenon is useful in that it gives information about the surface topography, but, on the other, it is difficult in practice to obtain surfaces which are sufficiently flat to give clear streaked diffraction patterns like those in Fig. 3.10.

The second factor is associated with uncertainties in the wave vector of the incident electrons due to the finite convergence and finite energy spread in electron beams coming from real sources. These uncertainties reveal themselves as a region of finite extent over which the electrons can be regarded as being in phase with each other. This region is called the *coherence zone*, and an expression for its size is derived in the caption to Fig. 3.11. In a typical RHEED experiment at 100 keV, the energy spread may be 0.5 eV and the convergence angle 10^{-5} rad. In these circumstances, time incoherence is negligible and the spatial incoherence zone is about 200 nm in diameter. Ordered regions on the surface very much smaller than this will broaden the diffracted beam to the point where it will not be observable. If they are large compared to 200 nm then sharp well-defined diffraction patterns will be

(a)

(b)

Fig. 3.10   RHEED patterns obtained at 100 keV from the (111) surface of silicon. (a) [211] azimuth; (b) [101] azimuth. The streaking indicates that the surface is flat.

obtained. A RHEED pattern of half a monolayer of oxygen on Cu(110) is shown in Fig. 3.12.

Structure analysis by RHEED

Although the symmetry of the surface unit mesh and the dimensions of the mesh can be determined by RHEED, as described above, the structure within the unit mesh can only be determined by analysis of the intensities of the

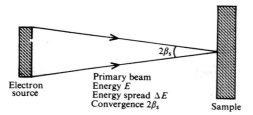

Fig. 3.11 The coherence zone diameter. See, for instance, Heidenreich (1964) for further explanation.

(a) Spread ΔE in energy of incident electrons gives *time incoherence:*

$$E = (\hbar^2/2m)k^2, \qquad \Delta E = (\hbar^2/2m)2k \, \Delta k,$$

therefore

$$\Delta k^t = k \, \Delta E/2E.$$

Resolved parallel to surface, for small β_s,

$$\Delta k_\parallel^t \approx (k \, \Delta E/E)\beta_s$$

(b) Spread in arrival angle over $2\beta_s$ gives *spatial incoherence:*

$$\Delta k_\parallel^s \approx k2\beta_s.$$

(c) Combine uncertainties in quadrature and define *coherence zone diameter* ΔX:

$$\Delta X \Delta k_\parallel = 2\Pi$$

$$\Delta k_\parallel = \{(\Delta k_\parallel^t)^2 + (\Delta k_\parallel^s)^2\}^{1/2}$$

$$\Delta X \approx \frac{\lambda}{2\beta_s\{1 + (\Delta E/2E)^2\}^{1/2}}$$

diffracted beams. This is exactly analogous to the X-ray diffraction case described above. However, unlike X-ray scattering, the cross-sections of atoms for elastic scattering of electrons are very large. This means that processes in which the ingoing electron beam may be elastically scattered several times before emerging from the sample are very probable. The theory of diffraction for a beam that has been scattered, say, three times interfering with another beam that has been scattered, say, seven times is very much more complicated than that for single scattering of all beams. This is multiple scattering theory, which has been most extensively developed for use in the interpretation of LEED experiments. This is described in more detail below.

Multiple scattering theories have been developed for RHEED intensity analysis and are of a comparable complexity and difficulty to those available for LEED intensities. The reader is referred to papers by Masud *et al.* (1976) and Maksym and Beeby (1984) for descriptions of these theoretical methods.

Unfortunately, there is a shortage of experimental data (compared with that for LEED) for comparison with theories of RHEED. This may be because the

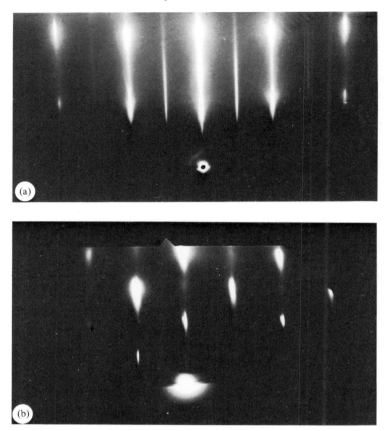

Fig. 3.12 RHEED at 100 keV from Cu(110)c(2 × 2)-O. The coverage corresponds to large flat islands of oxygen at a level of about half a monolayer. (a) Beam along [001]. (b) Beam along [130]. (By courtesy of F. Gronlund and P. E. Hojlund Nielson.)

experiment demands the use of very flat sample surfaces in order that the grazing incidence and emission can occur without transmission through scratches and bumps on the surface. Further, a very small diameter electron probe is essential because the projection of the beam on to the surface in the plane of incidence can be quite extensive. Thus the beam interacts with a large enough area of the surface that it may consist of several differently ordered regions. These problems are by no means insuperable, and it may be that structure analysis by RHEED will become more common.

RHEED intensity oscillations

Because RHEED experiments are conducted at grazing incidence the geometry is very convenient for the observation of diffraction processes while material is being deposited upon a surface. Indeed, the use of RHEED during the growth of thin layers by molecular beam epitaxy (MBE) has become a standard method. The processes of film growth and the MBE technique are described in Chapter 6. What is of interest here is that the intensity of RHEED beams can oscillate as material is deposited upon a flat single crystal surface. These oscillations are observed only when the depositing material grows monolayer by monolayer, rather than by nucleation and growth of small islands.

An example of RHEED oscillations is shown in Fig. 3.13. Here data are shown for monolayer growth of GaAs upon a GaAs(001) 2×4 surface. The growth is started at time $t = 0$ at the left-hand side of the plot. The oscillations are very clear and each period corresponds to the growth of one monolayer.

Early attempts to explain these oscillations proposed that elastically scattered electrons were interfering after being scattered from the top of an adsorbed monolayer and from the substrate. The situation is not as simple as this because it is observed that the phase of the intensity oscillations (see Fig. 3.13) varies with angle of incidence, there can be an initial transient intensity change when deposition is started and there can be double period oscillations in some diffraction geometries. The inelastically scattered electrons appear to contribute to the measured intensities, and so these must be included in any theory of the oscillations. The reader is referred to the article by Joyce *et al.* (1986). Oscillations are also observed in diffracted X-ray intensities as the coverage is increased monolayer by monolayer.

Low-energy electron diffraction (LEED)

The high atomic scattering cross-sections for electrons with energies less than 1000 eV (Fig. 3.4) suggest that low-energy electron diffraction should be extremely sensitive to surface atomic arrangements. This has been known for a very long time—indeed, Davisson and Germer used LEED in 1927 to demonstrate the wave nature of the electron. However, it is only since the development of UHV technology that LEED has been widely studied and used.

The Ewald sphere construction of Fig. 3.8 can be used to describe the electron beams that will be diffracted from a single-crystal surface. If only the top monolayer of atoms is responsible for the scattering, then a beam with wave vector $\mathbf{k}_0$, incident at an angle ϕ will produce backscattered beams $\mathbf{k}_1'$, $\mathbf{k}_2'$, $\mathbf{k}_3'$, and $\mathbf{k}_4'$ returning to the vacuum and forward-scattered beams $\mathbf{k}_5'$, $\mathbf{k}_6'$, and $\mathbf{k}_7'$ going on into the solid. By varying the primary energy (and so varying the radius $|\mathbf{k}_0|$ in Fig. 3.8) the number and directions of the scattered beams will vary. The single crystal surface will thus produce a spot diffraction pattern which will contract towards the specularly reflected beam ($\mathbf{k}_1'$ in Fig. 3.8) as the primary energy is increased. If this diffraction pattern can be observed and the

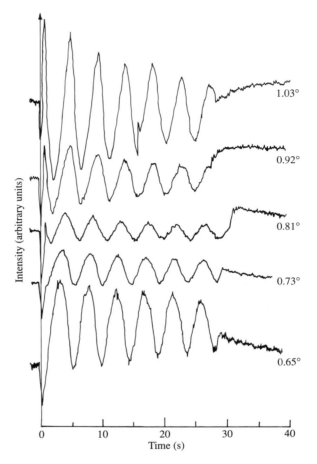

Fig. 3.13 RHEED intensity oscillations observed at five different grazing angles as GaAs is deposited upon a GaAs(100)2 × 4 surface. The incident beam has energy 12.5 keV and is in the [110] azimuth. The intensity of the specular spot in the (00) reciprocal lattice rod is being measured. Note that the period of the oscillations is independent of the grazing angle, but that the phase of the oscillations does depend on this angle. (By courtesy of Dr P. J. Dobson.)

angles 2θ between the scattered beams and the incident beam measured then the surface unit mesh can be determined using Fig. 3.8 (which is drawn for a square mesh). Geometrical consideration (using the same arguments as those under Fig. 3.2) of this figure gives

$$a = \frac{(h^2 + k^2)^{1/2}}{2 \sin\theta \, \cos(\theta + \phi)} \tag{3.15}$$

for any particular reciprocal lattice rod (hk). In the simpler case of normal incidence ($\phi = 0$), eqn (3.15) becomes

$$a = \frac{(h^2 + k^2)^{1/2}}{\sin 2\theta} \qquad (3.16)$$

and the slope of a plot of sin 2θ versus λ ($\lambda \approx 0.1 \, (150/V)^{1/2}$ nm) can be used to find the mesh side a. When $\phi = 0$, Fig. 3.8 shows how a simple symmetrical LEED pattern will be observed for normally incident electrons.

A LEED pattern can be displayed by using the arrangement shown in Fig. 3.14, which is the same electron optical arrangement used for the retarding field analyser in Auger electron spectroscopy (Fig. 2.3(b)). A beam with small convergence ($2\beta_s \sim 0.01$ rad, Fig. 3.11) and variable energy is diffracted from the crystal surface and backscattered beams move through field-free space between the crystal and grid G1. Between G1 and the screen they are radially accelerated so that they are energetic enough to excite fluorescence in the screen S and the spots of light so created are viewed or photographed through the window. The grids G2 and G3 are provided so that those electrons that have been scattered inelastically in the sample can be rejected and mainly elastically scattered electrons reach the screen. The potential V in Fig. 3.14 would be adjusted so as to reject these inelastic electrons, since they contribute only to a diffuse background in the diffraction pattern. Two grids are used to help reduce the effects of penetration of the high positive potential on the screen through the open mesh of the grids. This potential penetration simply reduces the quality of the arrangement as an electron energy analyser.

The means of holding the sample at the centre of the grid system is the subject of considerable experimental ingenuity. If the diffraction geometry is to be varied then both the angle of incidence and the azimuthal angle need to be adjustable; exploration of different parts of the surface requires two translations and adjustment of the surface to be at the grid centre requires a third translation; a study of diffraction intensities as a function of temperature (Chapter 5) may require both cooling and heating; cleaning up the sample surface may require heating to near the melting point of the sample or cleavage to produce a natural crystal face *in situ*. At the same time, the observer's view of the diffraction pattern is blocked by any experimental arrangement around the sample, and consequently any of these facilities must be provided in the most unobtrusive way possible. In some systems these difficulties are circumvented by depositing the fluorescent screen material upon a glass spherical section. The diffraction pattern can then be viewed in transmission through the screen and its glass support, but past the obstruction of the electron gun, which is made as small as possible. Such a construction is usually referred to as a *reverse view LEED system*.

The symmetry of the atomic arrangement at a surface and the sides of the unit mesh (eqn (3.16)) can be obtained immediately in a LEED experiment without varying the diffraction geometry, as is necessary in RHEED. This is an important use of LEED because the progress of an experiment which

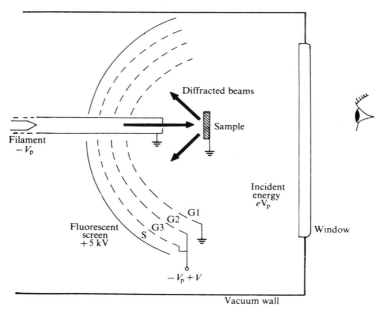

Fig. 3.14 LEED optics for displaying diffraction patterns. The screen S and the grids G1, G2, and G3 are spherical sections with a common centre at the point where the incident electron beam strikes the sample surface. The incident beam is usually focused to about 0.1–1 mm diameter at the sample, has a maximum current of about 2 μA, and an energy between 10 eV and 1000 eV. With different electrical connections the same electron optics can be used as a retarding potential analyser for electron spectrometry (Chapter 2).

changes the surface in some way can be quickly and conveniently followed. Some examples of LEED patterns from three types of surfaces are shown in Fig. 3.15, alongside indexed sketches of the same patterns. Figure 3.15(a) is typical of most low-index metal surfaces after cleaning, in that it shows the diffraction pattern expected from the bulk exposed plane of atoms. The clean surfaces of semiconducting materials are different from those of metals because they often give diffraction patterns indicating reconstruction (e.g. Fig. 3.15(c)). This is probably because of the directional character of the covalent bonds in these materials.†

In order to establish the surface sensitivity of AES, advantage could be taken of the fact that some metals grow upon others monolayer by monolayer (Chapter 2). The same fact can be used to examine the surface sensitivity of LEED. For example, the systems Au(100)–Ag, Cu(100)–Ag, W(100)–Cu, and Cu(100)–Au all show this monolayer-growth process and the LEED features

---

†For further discussion of LEED patterns and their explanation see the review paper by Jona *et al.* (1982).

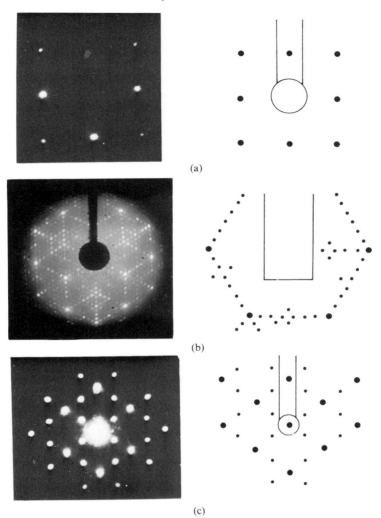

(a)

(b)

(c)

Fig. 3.15 Some typical LEED patterns. (a) The clean Cu(100) surface. Only spots from the bulk exposed plane are present. Primary energy 150 eV. (By courtesy of Dr R. J. Reid, New University of Ulster.) (b) The clean Si(111) surface. Extra spots are present between the six (10) features (bright spots) from the bulk exposed plane. These extra features correspond to a surface mesh parallel to the substrate mesh but with 7 times the length of its sides. Because of these 1/7th order spots this pattern is called Si(111)(7 × 7) or Si(111)7. Primary energy 42 eV. (c) The W(110)-(2 × 1)-0 LEED pattern due to oxygen adsorbed on W(110). Primary energy 53 eV. (By courtesy of Dr J. W. May, Eastman-Kodak Laboratories, New York.)

of the substrate are no longer detectable after between one to six monolayers of the deposit and primary energies below 125 eV. The precise sensitivity depends upon the particular substrate–deposit combination. This kind of result has to be treated with care if the deposit does not grow with a monolayer habit. Using Fig. 3.11 and the parameters of typical LEED systems ($2\beta_s \sim 0.01$ rad, $\Delta E = 0.5$ eV) then the coherence zone diameter is about 10 nm at a primary energy of 100 eV. Should a deposit consist of oriented islands greater than 10 nm in size then a LEED pattern will be seen which is difficult to distinguish from that due to a continuous oriented deposit.

By combining LEED and RHEED in the same apparatus the different diffraction geometries and coherence zone diameters can be used to draw conclusions about the surface topography and the extent of the ordered coverage as well as the size and symmetry of the surface mesh.

The high surface sensitivity of LEED arises from the large scattering cross-sections of atoms for low-energy electrons. This very property, which makes LEED so useful for surface information, also results in difficulties in the theoretical interpretation. Because the cross-section is high for elastic scattering at a single atom, it is possible for beams to be scattered several times and still emerge from the surface with measurable intensity (Fig. 3.16). The theory of X-ray diffraction, upon which rest the methods described on pp. 49–52, is a *kinematical theory*, in which the probability of such multiple scattering effects is treated as negligible. If multiple scattering occurs then all the waves scattered into a particular direction in many different scattering sequences must be added up with due regard for their correct amplitudes and phases. Such a treatment is referred to as a *dynamical theory*, and it is essential in the description of LEED.

The occurrence of multiple scattering can be demonstrated in practice by two kinds of experiments. In the first the intensity of a particular diffraction spot is measured as a function of the primary energy. Such a plot is usually referred to as an $I(V)$ plot, and an example is shown in Fig. 3.17(a) for MgO(100). If normal three-dimensional diffraction had been occurring then a maximum intensity should occur whenever the incident wavelength had the correct value for Bragg's equation (Fig. 3.2) to be satisfied. The so-called 'Bragg peaks' are indicated in Fig. 3.17(a). It can be seen that extra peaks occur in the $I(V)$ plot. These are sometimes called secondary peaks and are due to the multiple scattering. In the second experiment the primary energy is held constant and the crystal is rotated about its surface normal while the intensity of the (00) beam is measured. The graph of intensity I versus azimuthal angle ψ is called *Renninger plot* or *rotation diagram*, and an example is shown in Fig. 3.17(b) for W(110). Looking at Fig. 3.8, it can be seen that such a rotation does not change the angle of incidence ϕ, and a kinematical theory would suggest that $I(\psi)$ should be independent of ψ. However, if multiple scattering occurs then every time the value of ψ is such that strong diffraction can occur in some direction other than $\mathbf{k}'_1$, intensity must be lost from the beam $\mathbf{k}'_1$. Thus, the rotation diagram has minima everywhere the diffraction geometry is correct for strong scattering in some other direction from that being observed.

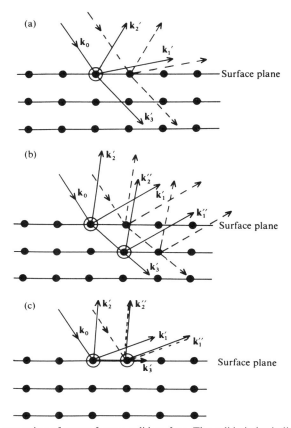

Fig. 3.16 The scattering of waves from a solid surface. The solid circles indicate the atomic positions and the open circles are added when a scattering event occurs. (a) Single scattering. Kinematical theory would describe such an event. Diffraction occurs when the 'dashed' wave adds up in phase with the 'solid' wave. (b) Double scattering. The forward-scattered beam k_3' is scattered into two beams k_1'' and k_2'' adding to beams k_1' and k_2' generated at the first scattering event. (c) A case of double scattering involving a surface wave k_3'.

Structure analysis by LEED

Theoretical studies of LEED and RHEED usually have one of two objectives. They may be attempts to explain the measured intensities in $I(V)$ or $I(\psi)$ plots in LEED or in $I(\phi)$ plots in RHEED in terms of the properties of the solid—the kinds and positions of atoms present and the potential in which they are situated. Alternatively, they may be attempts to determine the atomic structure from the measured intensities of spots in patterns like those shown in Fig. 3.15. Of course, both objectives have a great deal in common, but the

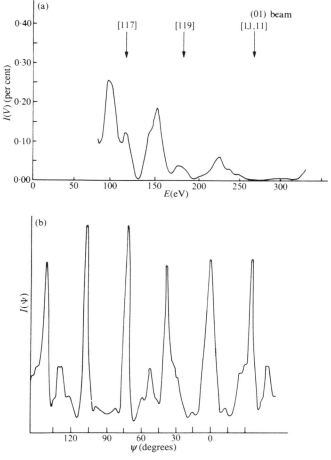

Fig. 3.17   (a) An $I(V)$ plot from MgO(100) in normal incidence. The energies of the Bragg peaks are marked and indexed. The shift between the large observed peaks and the calculated Bragg peak positions is due to the 'inner potential' in the crystal. (b) A rotation diagram $I(\psi)$ from W(110). The primary energy is 595 eV and the diffraction geometry is set so that the (550) beam is being measured. (After Gervais *et al.* 1968.)

motivations, and therefore the parameters which are treated as adjustable, are very different.

Although measurements of the geometry of the diffraction pattern enable conclusions to be drawn about the size and symmetry of the unit mesh, it is only possible to discover where each kind of atom is situated within the mesh and where it is placed above the next layer of atoms by measuring the intensity of diffraction features as the diffraction parameters are varied. The procedure for structure analysis is thus very similar to that described in Fig. 3.1, except that the difficulty of computing I_{hk} and refining the guessed structure after comparison with the experimental data is very much greater than it is in the X-ray counterpart.

The theoretical situation in LEED is eased slightly by the fact that the mean free path before inelastic scattering is quite small (a fraction of a nanometre in this energy range, see Fig. 2.7, p. 24). This means that there is a restriction on the number of elastic scattering events that can occur before inelastic scattering spoils the coherence of the beam. This can be seen from Fig. 3.16, which indicates how the path length in the crystal of the elastically scattered beams increases as the number of scattering events increases. Thus a multiple scattering theory in LEED does not require too many scattering events to obtain reasonable agreement with experiment. Nevertheless the theory of LEED is still very complex (Pendry 1974).

There are several steps in the procedure for calculating an $I(V)$ curve from a model of the structure which is a proposed solution to the problem of finding the atomic arrangement. The first set of steps are those involved in computing the electrostatic potential of the atoms in the model structure—this is needed because it is this potential which scatters electrons. The input to these first steps might be tables of the radial dependence of the electronic wave functions of each kind of atom involved in the structure. The potential can be assumed to be spherically symmetric, out to a certain radius, around the nucleus of each atom and then to be constant between these spheres—a so-called '*muffin tin*' potential. It is relatively straightforward to compute the charge density around each atom from the given wave functions and then to compute the muffin tin potential from these charge densities. This is a rather crude approximation to the actual potential in a crystal because it is a simple summation of the potentials of independent ions or atoms and it neglects the interactions between them. These interactions have the greatest effects in modifying the states of the outer electrons of each atom. Fortunately, in the LEED experiment the observation is of the intensities of electron waves *backscattered* from the crystal. The large scattering angles involved in the experiment are attributable mainly to the potential in the core regions of the scattering atoms or ions and the valence electrons have only a small influence on the backscattered intensities. Consequently, the relatively crude approximation of a muffin tin potential turns out to be acceptably accurate.

By exploiting the spherical symmetry of each muffin tin sphere, it can be shown (see Pendry 1974) that the scattering of electrons by each atom can be described by a set of phase shifts to the components of the incident plane wave expanded spherically into 'partial waves'. These phase shifts are dependent upon the energy of the incoming electron and can be calculated from the potential.

The next step begins on the task of calculating the scattering of the electrons by the array of atoms in the crystal. One convenient way of carrying out this summation is to regard the crystal as being made up of layers of atoms in sheets parallel to the surface. A simple crystal may be made up of a set of identical layers with neighbouring layers related by a translation parallel to the surface. More complex crystals may have several layer types, in addition to the translations. The task of calculating the effects of multiple scattering can then be split into two parts. The first is to compute the amplitudes and phases of all beams, leaving one layer for any incident plane wave. This should include all multiple scattering between the atoms of the layer. The result of this calculation is a reflection and a transmission matrix for the layer. In the second stage the layers must be combined to make up the crystal and the beams allowed to propagate between the layers being reflected and transmitted as they progress inwards or outwards.

One way of adding the beams together as they progress through the crystal makes use of the theoretical and experimental observation that forward scattering is usually much stronger than backscattering in the LEED range of energies. A perturbation scheme can be constructed based upon this fact and the additional fact that some of the electrons are scattered in inelastic events in the crystal and so are lost to a measurement of the intensities of elastically scattered beams. Such a perturbation scheme, called *renormalized forward scattering* theory, has been devised by Pendry (1974) and is outlined in Fig. 3.18. At the top of the figure an incident plane wave is

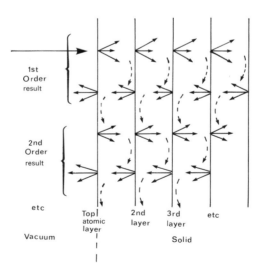

Fig. 3.18 Sequential 'passes' of the electron scattering processes involved as a renormalized forward-scattering scheme is used to compute the intensities of beams leaving a surface. Convergence is usually obtained after a few passes.

shown approaching the surface layer from the left. The forward scattered beams generated by each layer can then be recorded as they propagate into the solid until inelastic excitations and the elastic backscattering at each layer cause them to die out. Then, starting at the deepest layer reached in this first stage, the waves travelling back towards the surface are traced, adding the reflections from each layer due to the initial waves at each step. The end of this first 'pass' is a set of plane waves emerging from the crystal. A second pass can now be made starting with the emerging plane waves from the first pass and reflecting them at the surface layer so that they propagate back into the crystal. Now they are transmitted through each layer while adding to the reflected waves on their way. All these additions are, of course, done with proper regard to both the amplitudes and phases of all the waves. These passes are continued until the total reflected amplitudes have converged to within some specified accuracy.

In this kind of approach the inelastic scattering is regarded as occurring between the spherically symmetric ion cores in the region where the potential is assumed to be uniform. This is reasonable because one of the most probable inelastic scattering events in the LEED energy range is plasmon excitation (see Chapter 4). Since the conduction or valence band electrons participating in this excitation are not nearly so localized as the ion core electrons (which participate in the elastic backscattering) this is a reasonable model.

The electron–electron interaction in the region between the ion cores also has a part associated with exchange correlation effects. The net effect is that an *optical potential* can be used in this region:

$$V_o = V_{or} + i V_{oi}. \tag{3.17}$$

The real part, V_{or}, includes some of the exchange interaction of the incoming electron with the electrons in the solid and accounts for the refraction of an incident beam as it passes from the electron optically 'rare' vacuum to the electron optically 'dense' solid. The imaginary part, V_{oi}, includes the inelastic excitations which cause loss of electron flux from elastic scattering events. Usually, V_{or} is about -10 eV and V_{oi} about -5 eV.

As this lengthy description may indicate, the computer codes needed to carry out these calculations can be very large. Typically, calculations might use 9–11 energy-dependent phase shifts to describe the scattering potential and require that 50–100 beams be propagated into the crystal for adequate convergence in the emergent amplitudes. The whole calculation may occupy many megabytes of memory in a computer and use several hours of computing time to test several tens of different structural models of a surface. So-called 'supercomputers' (array or vector processors) have been very useful for this kind of calculation.

An example of the output of such a calculation is shown in Fig. 3.19, where the calculated intensities for the (12) beam from a clean Cu(111) surface are compared with experiment. The theoretical curves are for five different

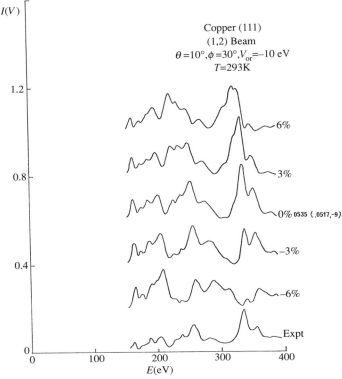

Fig. 3.19   LEED *I*(*V*) curves for the (11) beam from a Cu(111) surface. The geometry is for angle of incidence of 10° and azimuth of 30°. The bottom curve is the experimental data. The five top curves are for top layer relaxations of ± 6, ± 3, and 0 per cent of an interlayer spacing of bulk Cu.

distances between the top layer of atoms and the second layer (the relaxation) and the sensitivity of the calculated *I*(*V*) curves to the atomic spacing can be seen. The best agreement between theory and experiment in this figure can be seen for the zero relaxation theory. The solution of the structure may involve comparison of tens of such experimental *I*(*V*) curves with calculations for several structural models, each with at least one adjustable structural parameter, such as the relaxation of the surface layer of atoms. Naturally, numerical methods are used to make this comparison and choose the structural model which gives the closest agreement with the experimental data.

Hundreds of structures have now been solved using this technique. They include the low-index faces of single crystals of most elemental metals, the (100) faces of some metal oxides, some semiconducting compound surfaces, and a considerable number of ordered adsorbates, such as Cl, I, S, Se, Na, and

CO on low-index metal surfaces. Some of the results of this work are referred to in the book by Van Hove and Tong (1979) and the compilation by Watson (1987).

In RHEED the experimental inconvenience of varying the primary energy means that intensity data is usually in the form of rocking curves ($I(\phi)$) or rotation diagrams ($I(\psi)$). Although the scattering cross-sections are smaller in RHEED then in LEED, and therefore multiple scattering might be expected to be less important, the mean free path before inelastic scattering is much longer. This has the effect that more scattered beams must be included for an accurate description of the intensity. The net effect of smaller scattering cross-sections and longer mean free paths is to lead to a similar theoretical difficulty.

Search strategies and tensor LEED

The computational demands for surface crystallography by LEED can be very large indeed. Consider, for instance, that a surface containing two atoms per unit mesh is suspected to be reconstructed in the top three atomic layers. If the interatomic separations are completely uncorrelated (not necessarily the case, but this is the worst possibility) then three coordinates are required to specify the position of each atom. Thus, each layer of the solid requires the determination of six atomic coordinates and the whole surface requires the determination of 18 atomic coordinates. The LEED calculation needs to be carried out for each combination of the 18 coordinates and compared with the experimental data. Some measure of the degree of agreement between calculated and experimental $I(V)$ curves is required, and this measure must be used to determine the set of atomic coordinates which gives the best theory/experiment agreement. This measure is called an *R factor*, after a similar parameter used in X-ray diffraction studies of bulk crystallography. Normally the *R* factor is designed such that it passes through a minimum value at the best theory/experiment agreement. The 18 atomic coordinates and the *R* factor together define a 19-dimensional space which must be searched for a minimum in the *R* factor. Worse, this space is like a multidimensional mountain range, containing peaks, valleys, bowls, and cols. The height of each point in the mountain range is determined by a dynamical LEED calculation (computationally intensive) followed by an *R* factor calculation (relatively quick and easy). The problem is to find the minimum height in a parameter space bounded by the extreme values of all of the atomic coordinates. This is a problem of non-linear optimization—a field of study in itself!

Because of the computational cost of exploring such multidimensional spaces, there are many examples in the literature where physical insights or postulated surface structures have been used to limit the volume of parameter space used to find what may turn out to be a local minimum in the *R* factor. In these circumstances it is not surprising to find disputes in the literature about which structure accounts best for the experimental observations. Nevertheless, it should be stressed that LEED crystallography provides a critical test of a postulated structure. If the $I(V)$ curves calculated for the postulated structure do not correspond closely to the measured $I(V)$ curves then either the postulate needs small adjustments or it is wrong. This is sometimes referred to as the *LEED test* of the postulated structure.

One way of reducing the computational cost is to employ a strategy for searching the parameter space which uses full dynamical calculations only when absolutely necessary. A first step might be to carry out a coarse exploration of the space using relatively few electron energies to calculate a set of points on each $I(V)$ curve for each rather coarsely spaced choice of each atomic coordinate. This set of calculations gives a basis set which can then be searched in more detail. A local minimum in the basis set is chosen as a base point in parameter space (the initial structural model) from which exploratory moves are made by making a step either side of the base point in each dimension of the parameter space. Comparison of these exploratory models with the experimental data, using R factors, may lead to a lower R factor and hence a new base point from whch exploratory moves are again made. The search around the original base point is stopped when exploratory moves, using sufficiently small increments in each parameter, provide no further improvement in the agreement with the experimental data. A complete dynamical calculation at the 'best' point near to the original base point can then be carried out. There may be other minima in the original basis set and each of these must be used in turn to explore for the local minimum in the R factor. Finally, the deepest minimum can be located and the parameters here provide LEED $I(V)$ curves which are the best fit with experimental results.

This approach has been used for example to resolve a dispute over the crystallography of the (110) surface of III–V group semiconducting compounds (Cowell and De Carvalho 1988). They report on the use of an experimental database of 23 symmetrically inequivalent $I(V)$ curves collected in both normal and off-normal diffraction geometries. These data were compared with 20 different structural models of the CdTe(110) surface with a search strategy in 11-dimensional data space. Their results are summarized in Fig. 3.20. The parameters characterizing the reconstruction of the CdTe(110) surface are indicated in Fig. 3.20(a). The bulk exposed plane would have coplanar Cd and Te atoms without relaxation ε_1 or rumpling ρ_1. The results

---

Fig. 3.20 The structure analysis of CdTe(110)—a cleavage surface of this compound. (a) indicates how the bulk structure (on the left) is modified at the surface (on the right). The symbols indicate the parameters which are sought in a determination of the surface crystallography. The R-factor plot (c) shows how several minima of different depths can occur in the possible multidimensional parameter space being searched. This particular diagram contains a cut through this space revealing the dependence of the R-factor upon the two structural parameters which affect it most strongly. The $I(V)$ curves (b) compare experiment and theory for the best agreement found.

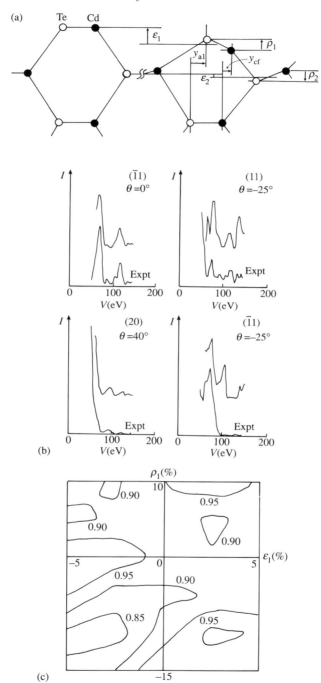

are presented as a contour plot if the R factor, which shows its deepest minimum where experiment agrees best with theory. This is shown in Fig. 3.20(c). This work suggests that the surface layer is relaxed towards the bulk by 3.7 per cent of a bulk lattice constant and the line between the Cd and Te atoms in the surface has rotated away from the surface plane by 17°. Even the second layer atoms have moved from their bulk positions by about 1 per cent of the bulk lattice constant.

Another development of LEED theory is particularly appropriate for use together with this search strategy, but is, in any case, a faster way of exploring the parameter space. This is called tensor LEED which was invented by Rous *et al.* (1986) and is described by Rous and others (1991). The idea here is to calculate first the $I(V)$ curves for a *reference structure* using a full dynamical theory as outlined above. This structure is chosen to be as near to the expected surface structure as possible. Subsequently, a whole set of *trial structures* are examined by making small variations in the atomic positions of the reference structure. The $I(V)$ curves of the trial structures can be computed very quickly by using a perturbation scheme in which the differences between the amplitudes of the scattered electron beams from the reference and a trial structure are expanded as functions of the atomic displacements which generate the trial structure. To first order this a linear expansion in the displacements and the weights by which they are combined to form an amplitude difference form a tensor—hence the name of the method. The substantial advantage of this approach is that a whole region of a multi-dimensional parameter space can be searched quickly after just one computationally expensive multiple scattering calculation. Greater precision can be obtained by using a non-linear expansion.

By combining the tensor LEED approach with a search strategy, as described above, it is possible to explore a huge volume of parameter space and find a global minimum in the R factor with relatively few full multiple scattering calculations. Of course, it is essential to start the search for trial structures at several places in parameter space in order to avoid finding what may turn out to be a local rather than a global minimum. Also, the whole process is completed with a full dynamical calculation at the end of the search. Rous *et al.* find that they can expand the displacements in such a way that trial structures with atomic displacements up to 0.04 nm away from the reference structure can be evaluated and that improvements in the speed of the whole structural determination of the order of 50 times can be obtained.

Surface defects

All the discussion of LEED to this point has been concentrated upon determination of the surface crystallography of an ideal single crystal surface. Any real surface contains defects, which depend for their type and number upon the temperature of the surface and its history—it may for example have been etched or ion bombarded or stressed to near breaking point in some previous experiment. Henzler (1984) has shown how the analysis of the variation of intensity across a LEED beam can be used to extract valuable information about these defects. This spot profile analysis he has dubbed *SPA-LEED*.

As described above, a perfect periodic arrangement of atoms in a surface leads to a spot diffraction pattern in LEED. The arrangement of spots is given by the periodicity of the surface and the sharpness of the spots by instrumental factors—the angular and energy spread of the incident electron beam and the energy resolution of the electron optics of the LEED grids. Deviations from the diffraction pattern of such an ideal surface contain information about the imperfections in the surface. Examples of such imperfections are point defects, static or dynamic disorder, steps, and domains with different orientations in the surface.

The profile of the spots observed from a single crystal surface can reveal, for example, splitting and broadening. Splitting can occur when there is a periodicity in the occurrence of some defect like steps in the surface. If the steps have a periodicity of 20 times the size of the unit mesh, then spitting of spots will be observed which is 1/20th of the length of the side of the diffraction pattern, and this will be in the direction normal to the step length, i.e. in the direction of the periodicity. Broadening occurs when there is a distribution of periodicities about some mean value—the wider the distribution the greater the broadening of the diffracted beams. An example of such a case is provided by epitaxial growth (see Chapter 6) in which ordered islands of material can grow in a single crystal substrate with a distribution of distances between neighbouring islands. SPA-LEED can be used in the case to determine the mean distances between islands and to examine, for example, how this depends upon the rate of arrival of atoms in the growth and the temperature of the surface. The complementary observations of spot profiles and STM images (see later in this chapter) should be very powerful in this kind of study.

Low-energy electron microscopy (LEEM)

A microscope using the diffraction of low-energy electrons in a surface was invented by Bauer (1962) and has proved to be powerful in revealing the details of processes occurring as a single crystal surface is changed by heating and/or deposition of various kinds of atoms. Figure 3.21 shows the arrangement of components in this microscope. A pointed cathode in an electron gun is used to produce a 20 keV electron beam which can be focused using a conventional magnetic column of the kind typically found in transmission electron microscopes (TEMs). The focused beam is deflected into a special lens (a *cathode lens*) immediately in front of the surface to be studied, where it is retarded to an energy of between a few tenths of an eV to 150 eV. The low energy beam can then be diffracted by a single crystal surface

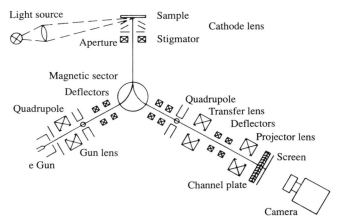

Fig. 3.21 The arrangement of a low-energy electron microscope. (By courtesy of Prof. E. Bauer.) The diffraction pattern can be excited either with a low energy electron beam formed in the cathode lens or by photoelectrons generated with a light source.

and beams emerge from the sample, just as they do in LEED. These electrons pass back through the cathode lens and are deflected in the magnetic sector, having been accelerated back to 20 keV in the lens. If the magnetic transfer lens is focused upon the back focal plane of the cathode lens, then an image of the LEED pattern can be formed upon the front surface of the micro-channel plate detector. Alternatively, the aperture in the back focal plane of the cathode lens can be positioned so as to allow a single diffracted beam to pass through the transfer lens, whose strength is adjusted to focus the image plane of the cathode lens on to the front surface of the micro-channel plates. If the aperture allows only the specular (00) LEED beam to pass through and form an image of the surface the process is called *bright field microscopy*. If the aperture is positioned to allow one of the diffracted beams to form an image this is called *dark field microscopy*. These two modes are analogous to those with the same names used in conventional (high-energy) transmission electron microscopy.

If the elastic scattering cross-section, the lattice constant, or the sample orientation varies from place to place in the surface then the intensities of the LEED beams will also vary from place to place by exactly the same arguments given above for the explanation of $I(V)$ curves. In turn, the brightness of the image in the dark or bright field modes will also vary from place to place.

Some particularly beautiful results have been obtained by Bauer and his co-workers when using LEEM for the observation of Si(111) surfaces subjected to a variety of heat treatments. At temperatures above about 1100 K the surface has the bulk Si lattice and so is described by the notation Si(111)1 × 1. below the transition temperature it undergoes a phase transformation to a reconstructed form wth a unit mesh seven times larger than the bulk unit mesh—Si(111)7 × 7. LEEM has been used to study this phase transition. Figure 3.22 shows how this transition nucleates at the atomic steps as triangular shaped regions of Si(111)7 × 7 are formed upon the Si(111)1 × 1 surface.

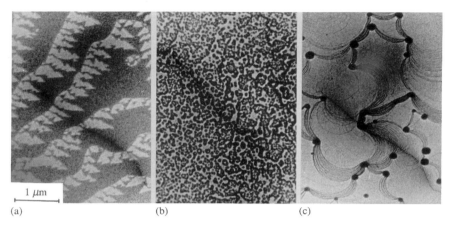

1 μm

(a)   (b)   (c)

Fig. 3.22 LEEM images of the Si(111) surface after various heat treatments. Electron energy 10 eV. (By courtesy of Prof. E. Bauer). (a) After heating to 1450 K and cooling to 5 K below the transition temperature for the (1 × 1) to (7 × 7) surface structure (ca. 1100 K); (b) after several hours heating at 1100–1150 K; (c) after prolonged heating at 1200–1300 K.

The same methods have been applied by Bauer and his group to the study of metallic surfaces and the epitaxial growth of other metals upon them. Some examples of the use of the technique are given in Bauer (1985, 1988).

Grazing incidence X-ray diffraction

Although it was stressed above that the cross-sections for scattering of X-rays are very small and that this means that the X-ray intensities diffracted by the surface atoms are small compared with those diffracted by the bulk material, nevertheless it is possible to obtain surface crystallographic information with X-rays. Two experimental steps are taken:

1. The sample is illuminated with a narrow pencil of monochromatic X-rays in grazing incidence. The angle between the incident wave vector and the surface needs to be very small to enhance the sensitivity of the diffraction experiment to the surface atoms. This is exactly analogous to RHEED as described above.

2. The intensity of the X-ray beam is raised as high as is possible provided that
   the sample is not damaged by the bombardment.

This kind of work is reviewed by Feidenhans'l (1989). Angles of incidence of
a few mrad are used, which leads to the necessity of preparing a very flat
sample surface. The high intensities required are usually obtained by using
synchrotron radiation, although the surfaces of the heavier elements have
sufficiently large scattering cross-sections for X-rays that high-powered (more
than 30 kW) conventional sources can be used. The geometry of the
experiment is indicated in Fig. 3.23 and a particular arrangement of
components used in the Hamburg Synchrotron Radiation Laboratory
(HASYLAB) is shown in Fig. 3.24. The diffraction experiment is mounted on a
beam line which is tangential to the orbits of electrons maintained in circular
trajectories around the synchrotron (Fig. 3.25). Because the electrons are

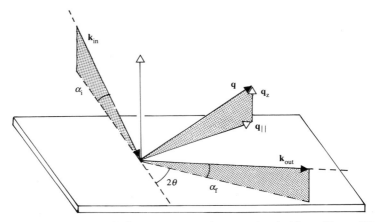

Fig. 3.23 The geometry and the wave vectors in a grazing incidence X-ray diffraction
experiment.

maintained at constant energy by the microwave magnetic fields and the static
beam bending magnets these machines are often referred to as storage rings.
The electrons are circulated at a well-defined energy for several hours. Since
they are being accelerated radially to keep them in a circular path they emit
Cerenkov radiation. Further, because the kinetic energies of the electrons are
high (2–10 GeV) their velocities are relativistic and the Cerenkov radiation is
emitted in the forward peaked direction along the tangents to the ring. Several
experimental stations can be mounted around a storage ring on a series of
beam lines. The principles and design of storage rings intended as bright, wide
wavelength range sources of light are reviewed by Wille (1991).

A very important property of synchrotron radiation is that it is polarized in
the plane of the electron orbit within the storage ring. This can be exploited to

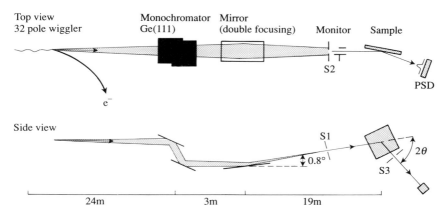

Fig. 3.24 X-ray wiggler beam line in the HASYLAB electron synchrotron. The monochromation of X-rays leaving the accelerator is performed by the pair of flat Ge(111) crystals. The toroidal double focusing mirror is Au plated. After diffraction the X-rays are detected by a position sensitive detector (PSD). (By courtesy of Prof. R. Feidenhans'l.)

place the electric field of the X-ray beam in a direction to optimize the intensities of the diffracted beams or to explore the electronic effects in the sample (see Chapter 4). In X-ray diffraction the sample is arranged to be in a vertical plane (the electron orbits are horizontal—after all, a storage ring is a very large machine!). With this geometry, reciprocal space can be explored in a horizontal plane with a movable single channel X-ray detector or with a fixed array of X-ray detectors (a *position-sensitive detector*). Because the reciprocal lattice is a set of rods normal to the surface, the sample can be rocked about a vertical axis in its plane or rotated about its surface normal to explore the diffracted intensities.

Scans of the intensity distribution along the reciprocal lattice rods corresponding to fractional beams from ordered adlayers of Pb or Sn upon Ge(111) are given by Fiedenhans'l (1984) and demonstrate the sensitivity of the X-ray method to surface atomic arrangements. The theoretical interpretation of the X-ray intensities to obtain crystallographic information is simpler than it is for LEED. This is because the only modifications necessary to simple kinematic scattering theory are for the absorption of the X-rays by the solid—dynamical scattering theory is not required. Several tens of surface structures have been solved by this method at the time of writing. However, the necessity of working with a surface which is very flat has limited the number of systems so studied.

Field-ion microscopy (FIM)

A technique with the ability to detect directly the positions of atoms upon a surface was invented by Muller (1951) and is called *field-ion microscopy*. It is

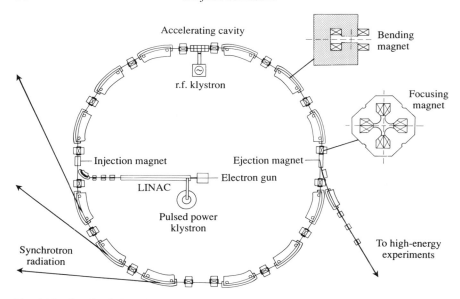

Fig. 3.25 Sketch of the principal components of a storage ring which emits synchrotron radiation by accelerating electrons on a circular trajectory at constant kinetic energy. Electrons are injected into the ring in pulses by a linear accelerator (LINAC). The bending and focusing magnets keep each pulse of electrons circulating around the ring. Some sections in the ring may contain special assemblies of periodically arranged magnets which accelerate the electrons on wavy trajectories with very small radii of curvature over a small region of space. This results in greater brilliance of the X-ray photon beam emitted along the tangents into the beam line, but it is localized to the position of the 'wiggler' or 'undulator'. (By courtesy of K. Wille 1991.)

reviewed in Muller (1965). A diagram of a field-ion microscope is given in Fig. 3.26. The specimen is prepared in the form of a sharp tip to which a positive potential is applied so that a field of the order of 5×10^8 V cm^{-1} is present at the tip surface. Molecules of the imaging gas (usually helium or neon at a pressure of $1–3 \times 10^{-3}$ Torr) move towards the tip and collide with it. After many collisons (Fig. 3.27) they are slowed down and lose an electron to the tip by quantum mechanical tunnelling. When this happens, the resultant positive gas ion is accelerated away from the tip in the large electrical field and strikes the fluorescent screen, causing a spot of light to be created.

According to one simple model, the ionization of the imaging gas atom is most likely to occur where the local electrical field is high, i.e. where the radius of curvature of the tip is highest. Thus a protruding atom is more likely to cause ionization of the imaging gas than a flat plane of atoms. The spatial resolution of the technique depends upon the component of the velocity of the imaging ions tangential to the tip surface. If the tip is cooled to reduce this component then resolutions of about 0.25 nm can be achieved. Therefore spots

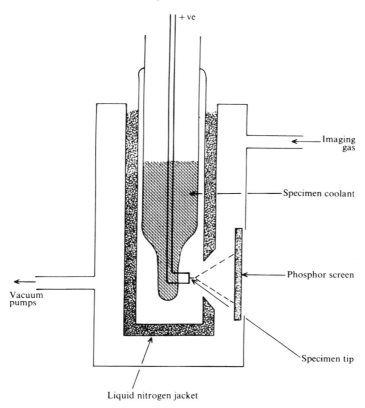

Fig. 3.26 A simple field-ion microscope arrangement. The specimen tip is mounted on a wire which can be heated by passing a current through it.

of light on the fluorescent screen correspond to the positions of individual atoms on the tip.

The material of the tip has to be a single crystal stable under the effect of high electrical field necessary to obtain ionization and has to be prepared so that needle tips having a radius of curvature well below 100 nm are possible. Tips of the elements W, Re, Ir, Pt, Mo, Ta, Nb, and Rh are all stable in the fields required to ionize helium, and the technique has been extended to tips of Zr, V, Pd, Ti, Fe, and Ni and some of their alloys with modifications to the imaging gas and the use of modern image intensifiers to help observation of the weak picture on the fluorescent screen. The tip is prepared by etching in the laboratory and then formed *in situ* in the microscope by raising the applied electric field until surface atoms are removed as ions. This field evaporation process is self-regulating, in that the field is greatest at sharp edges and aspherities and so these are removed first.

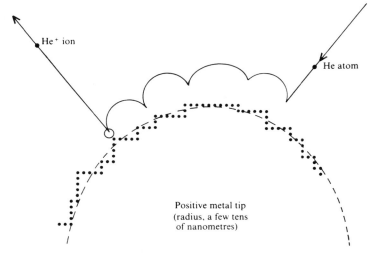

He⁺ ion

He atom

Positive metal tip
(radius, a few tens
of nanometres)

Fig. 3.27 A helium atom, polarized in the electric field, is drawn towards the tip. In a number of
hops it is slowed down until it is ionized in the region above a protruding atom. The He⁺ ion is
then accelerated away by the field towards the fluorescent screen. Real tips do not have such
simple atomic arrangements.

Figure 3.28 shows a field-ion micrograph of a platinum tip of 150 nm radius
imaged with helium. The spots correspond to the protruding atoms and those
circling the flat planes drawn in Fig. 3.27 can be clearly seen. Individual
vacancies, emerging dislocations, and regions of disorder can be detected with
the technique.

Because such a special simple geometry is required for FIM it is not possible
to use the technique for surface structure analysis in the sense of studying flat
single-crystal planes, as used in LEED. However, an enormous amount of
information about surface atomic arrangements can be derived for systems for
which a tip can be formed. Adsorbates can be studied on the tip if they are
sufficiently strongly bound not to desorb in the field. Some particularly
beautiful observations of the clustering and migration of small groups of
iridium atoms on W(110) surfaces are described in Chapter 6.

As mentioned in Chapter 2, FIM can be combined with a mass spectrometer
so as to make a device which can detect and identify an individual atom. The
specimen tip is arranged to be movable so that the image of a particular atom
falls on a small aperture in the fluorescent screen. A short pulse of voltage is
applied to the tip to cause field evaporation of the atom which passes through
the aperture, and its time of flight to a single-particle detector is measured. To
obtain statistically significant information on the local composition, many
atoms must be sampled, but the technique is one of outstanding sensitivity. It

Fig. 3.28 A field-ion micrograph from a platinum tip of about 150 nm radius imaged with helium at 28 kV. The centre of the image, for example, could be enlarged to reveal atomic scale detail in that region. (After Muller 1970.)

is referred to as the *atom-probe FIM* and is described in detail by Muller *et al.* (1968).

The field-electron microscope (FEM) is related to the FIM. It reveals variations in work function from place to place upon a tip and is described in Chapter 4.

Scanning tunnelling microscopy (STM)

All the methods for the determination of surface crystallography described above are scattering techniques. Because intensities of scattered waves are measured and their phases are unknown a theory is always required to invert the measurements in order to estimate the coordinates of the atomic positions. Because of the loss of phase information such an inversion is generally not directly possible, and procedures involving intelligent guessing of the structure, calculation of the intensities, and comparison of the experimental results with those from the guessed structure, as described above, have to be carried through. This indirect procedure is inconvenient and one is never sure that a different guessed structure would provide a better theory–experiment comparison than any found so far. It would be extremely valuable if a method could be found for somehow 'seeing' the atoms with a direct technique. Perhaps then the distances between neighbouring atoms could be measured directly on a photographic plate or something similar.

Such a technique was invented by Binnig and Rohrer in 1981 and was so successful and striking that they were awarded the Nobel Prize for it in 1986. The approach was so adventurous that experienced experimental physicists might have predicted that it would be impossible to do! Perhaps one should try first and predict afterwards? The subject has been reviewed by Guntherödt and Wiesendanger (1992) and by van de Leemput and Lekkerkerker (1992).

The principles of their *scanning tunnelling microscope* (STM) are shown in Fig. 3.29. A very sharp metallic tip mounted upon a set of piezoelectric transducers is supported very close to the surface of the sample to be studied. If a potential difference is applied between the tip and the surface and the tip–surface separation, t, is small enough then electrons can pass between them by quantum mechanical tunnelling. The tunnelling current is a very rapid (exponential) function of the separation t. The voltage across the z transducer controlling t can be arranged to be set by a feedback circuit whose input is the tunnelling current. The sense of the feedback is organized such that the tunnelling current is maintained to be constant as the other two (x and y) piezoelectric transducers are made to move the tip parallel to the surface. The voltages applied to the x and y transducers can be made, for instance, to cause the tip to move in discrete steps so as to scan over a square or rectangular area of the surface. Since the tunnelling current is kept constant by the feedback circuit then the voltage V_z applied to the z transducer is causing the tip to follow the microscopic topography of the surface, since it is maintaining a constant separation t. If the tip is atomically sharp then the variations in V_z reflect atomic scale variations in the topography of the surface! Such a microscope picture is usually called a *topographic image*. Approximately, the tip is following contours of constant charge density in the surface.

If the tip is not moved at all and the tunnelling current is measured as a function of the voltage beween tip and surface then information is obtainable

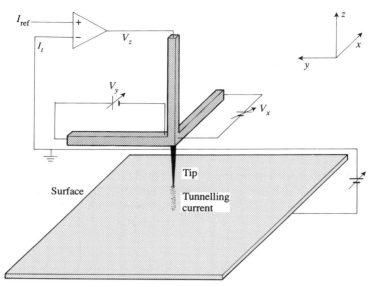

Fig. 3.29 The components of the scanning tunnelling microscope. The z piezoelectric transducer controls the separation of the tip from the surface. The x and y transducers scan the tip parallel to the surface.

about the densities of occupied and unoccupied states in the tip and in the surface. This is *inelastic tunnelling spectroscopy*, which is described in Chapter 4. If the surface is positive with respect to the tip then electrons tunnel out of the tip. The number that are free to leave the tip is determined by the density of occupied electronic states in the conduction band of the tip. However, they cannot form a current to the tip unless there are empty states in the electronic structure of the surface for them to go to. Therefore the current depends upon the density of occupied states in the tip, the density of unoccupied states in the surface, and the probability of crossing the intervening vacuum gap by tunnelling. Conversely, if the tip is positive with respect to the surface then electrons tunnel in the opposite direction. Now the current depends upon t, the density of unoccupied states in the tip, and the density of occupied states in the surface. Thus, different information is obtained depending upon the polarity of the tunnelling voltage. Because the tip is deliberately made to be atomically sharp the spectroscopic information so obtained is localized to the atom of the sample which is immediately under the tip.

Should the tip be scanned over the surface by the x and y transducers but now, at each tip position, the tunnelling voltage is varied, then a spectrum is measured at each point in a way depending upon atomically resolved densities of states in the surface. This is called *current imaging tunnelling spectroscopy* (CITS) and is discussed in Chapter 4.

There are three severe experimental challenges in making STM observations. First, the tip must be fabricated to be atomically sharp in order to obtain the spatial resolution necessary to 'see' individual atoms. Second, the mechanisms supporting the tip and the sample must be capable of achieving distances t of the order of 0.1 nm and maintaining them in a controllable way as the tip is scanned over the surface. Finally, the tip—sample assembly must be isolated from vibrations which would cause t to vary—either by coupling of motions of the laboratory floor or of vibrations in the air (sound) through the apparatus.

There are many recipes for the production of atomically sharp tips. The material of the tip is usually polycrystalline tungsten, and this can be etched using electrochemical methods to produce a point with about 100 nm radius of curvature. One way of further sharpening such a tip is to cause it to touch a corner of the sample in the STM and then withdraw it with the z transducer. The action of dragging the tip away from the bonding forces of the sample may cause an atomically sharp region to be left on the tip. Tips with a single atom on a protrusion can sometimes be formed in this way. Another method of sharpening is to apply a sufficiently large potential difference between tip and surface that atoms are field emitted from the tip. This changes the tip shape and sometimes an atomically sharp end contour can be achieved. None of these methods is reliable and repeatable, and so the formation of a 'good' sharp tip is something of a black art!

The piezoelectric transducers are a good solution to the means of holding and moving the tip. They are usually made of a barium titanium oxide material. A potential difference across a piece of such material causes a change in length proportional to the voltage applied. The voltage sensitivity is of the order of 10^{-10} m V^{-1} so that displacements of the order of 1 nm can be made with an applied potential difference of about 10 V.

Vibration isolation can be achieved by making the tip—sample assembly small and massive so that its natural vibrational modes have high frequencies; mounting the whole vacuum system containing the STM on anti-vibration mounts to reduce coupling of noise through the floor of the laboratory; situating the whole STM system in a soundproof or at least very quiet room; and mounting the tip—sample assembly on a critically damped anti-vibration suspension inside the UHV system. This can be provided, for instance, by suspending the tip—sample assembly on high compliance springs fixed at one end to the vacuum chamber. Damping of the motion of the suspended mass can be provided by mounting small permanent magnets on the tip—sample assembly and close to high conductivity copper components mounted rigidly on the vacuum wall. When the sample assembly moves eddy currents are induced in the copper which damp out the motion. Sometimes all of these anti-vibration precautions are included in the same STM system. Figure 3.30 shows a photograph of an STM with eddy current damping.

An example of a topographic image of the Si(111)7 × 7 surface is shown in

Fig. 3.30 (a) The tip–sample assembly of an Omicron STM in the author's laboratory. The tip is mounted on the part marked and the sample is mounted on the plate A in a plane perpendicular to the figure. The suspension rods and the eddy current damping system C can be discerned at the outside of the assembly. (By courtesy of Dr S. P. Tear, University of York.)

Fig. 3.31. The unit mesh of the 7×7 structure is marked in the image. It can be seen that there is a dark spot at each corner of the mesh surrounded by a characteristic 'rose' of six bright spots. This kind of image has been crucial in arriving at a model for the $Si(111)7 \times 7$ surface structure. The details are outlined in the case study near the end of Chapter 4.

Tunnelling can only occur between two objects if their wave functions overlap. The atomic scale resolution of the STM is obtained by confining the tunnelling to a very small area parallel to the surface. This is achieved by using a field-emission tip over the surface which is itself very localized by having a *very* sharp point—indeed, it is the goal of tip preparation to cause a single atom of tip material to be at the apex. This has clearly been achieved to obtain the STM image of Fig. 3.31.

STM studies of semiconductor surfaces are almost always technically 'easier' (relatively speaking) than those for metallic surfaces. This is because the wave functions of the valence band electrons of the surface atoms of a semiconductor are rather localized. (Hence the dangling bond idea for a semiconductor surface.) On the other hand, the conduction band electrons in a metal are generally free to move over very large distances—they are not localized. This is particularly the case for free electron-like metals and is less

40 Å

Fig. 3.31   Example topographic image. Si(111)7 × 7 taken at a tunnelling voltage of 2 V, 2 nA. The tip is negative with respect to the Si surface, so the contrast is modified by variations from place to place in the density of empty states in the Si. The distance from white to black adjacent dots is 1.5 Å. (By courtesy of Dr S. P. Tear.)

true for the transition metals, which have, for example, rather localized 3d electrons in their conduction bands. Therefore the STM measurements on metal surfaces require higher lateral and vertical resolution than they do for semiconducting surfaces. In turn this means that electronic and vibrational conditions need to be more stringently controlled and the tunnelling tip needs to be very sharp. Nevertheless, STM on metals has been successful and examples have been reviewed by Kuk (1992).

Strictly speaking, STM is not a technique which can be used for complete crystallographic specification of a surface structure. It does allow direct visualization of the arrangement of atoms in the unit mesh and the distances parallel to the surface can be estimated with precisions of the order of 0.01 nm by calibrating the displacement voltages applied to piezoelectric transducers against a known distance (like seven times the spacing of Si atoms in the Si(111)7 × 7 structure). However, it is not possible to determine the distances between the top layer of atoms and the underlying layer and the precision of determination of in-plane distances is about an order of magnitude lower than it is for LEED or grazing incidence X-ray diffraction. Nor can it be used to identify atomic species. Also, it should be noted that the height information in topographic images can be strongly influenced by the electronic structure of the surface (see Chapter 4). The great value of STM in surface structure determination is that it provides a direct view of the arrangement of surface atoms which can be used to eliminate many different structural models that might otherwise have to be modelled in a diffraction study. It is therefore a powerful complementary technique to the diffraction approaches.

Ion scattering

If a monochromatic beam of ions is scattered from a single crystal surface and a measurement is made of the number of ions scattered elastically into a detector as the emission direction is varied, then information about the surface structure can be obtained. This yield variation is caused, in part, by *shadowing* of the substrate atoms by adsorbate atoms (see Fig. 3.32). It can be seen from

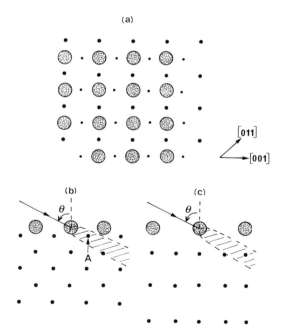

Fig. 3.32 Shadowing in low-energy ion scattering caused by the surface atoms. The figure is drawn for 1 keV Ar$^+$ ions incident upon a Ni(100) $\sqrt{2} \times \sqrt{2}$ R45°-O structure. (a) Plan view of the surface; (b) section through the (001) azimuth; (c) section through the (011) azimuth in which shadowing of nearest neighbour atoms to O atoms does not occur. (After Woodruff and Delchar 1989.)

this figure that, after scattering by a surface atom, an ion is not available for scattering by a nearest neighbour atom 'behind' the first atom. The extent of this shadowed region can be computed with knowledge of the energy and mass of the incident ion, the mass of the scattering atom and the ion–atom scattering potential. By varying the angle of incidence, θ, and the azimuthal orientation of the incident ion beam, the scattering due to the nearest neighbour atom varies according to whether it is in or out of the shadow. This variation can be intrepreted to define the site and spacings of the surface atom with respect to its neighbours. Unfortunately, these variations in yield are modified by processes in which the incoming ion can be neutralized by electron transfer from the solid. This charge neutralization varies with incident beam orientation and so must be included with the geometrical shadowing interpretation if adatom sites and spacings are to be determined. For incident ion energies below about 5 keV, charge neutralization is a significant effect and yet is insufficiently known for it to be adequately incorporated into the scattering theory. A more complete discussion of these effects can be found in Woodruff and Delchar (1989).

These drawbacks are not operative in high-energy ion scattering (>100 keV), where the dominant interaction between light ions and atoms in the solid is the internuclear Coulomb repulsion. The collisions involved in such an experiment obey the classical Rutherford scattering law. By studying the geometry of the shadowing effects in the yield of scattered ions (which have lost kinetic energy according to the masses of the atoms in the solid) it is possible to define surface atom sites and spacings. Furthermore, the theory of the interaction is sufficiently well known that in addition to the crystallography it is possible to identify the scattering atoms because their masses are determined from the energies of the scattered ions. This kind of analysis is common in the characterization of layered thin film structures and is usually referred to as *Rutherford backscattering* or *RBS*. An example of the determination of surface crystallography by these means is the reconstruction of the Pt(100) surface, which had been studied reported by Norton *et al.* (1979).

Angularly resolved photoelectron spectroscopy

As will be seen on p. 123 *et seq.*, it is possible to obtain a great deal of information about the electronic structure of surfaces by measuring and interpreting the angular and energy dependence of emitted photoelectrons originating from the valence and conduction band regions of the band structure. If the incident photon energy is raised to allow excitation of core level photoelectrons and the yield of these electrons is measured as function of their direction of emission, then information can be obtained about the surface atomic structure.

The interpretation of this method of surface crystallography requires the use of the LEED theory outlined above. In addition, the multiple scattering formalism has to be modified to treat the elastic scattering of the photoelectrons on their way out from a point source where they were generated inside the crystal (Fig. 3.33) through the periodic potential of the atomic arrangement to the vacuum. This approach to theory has been developed to study the emission angle dependence of both Auger electrons and photoelectrons, but it is more complex than the theory required for the incident plane waves of a LED experiment.

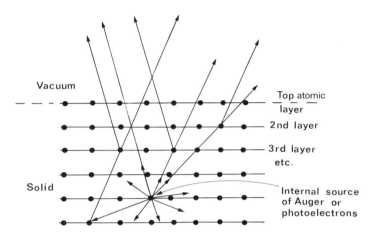

Fig. 3.33 Diffraction of electrons on their way out to the surface after leaving a spherically symmetric source (an emitting atom) somewhere inside the crystal.

This kind of experiment has been carried out by Fadley *et al.* (1979) using 1487 eV incident X-rays and the O 1s emission for a sample of Cu(001) with a c(2 × 2) overlayer of oxygen. The angularly resolved data are shown in Fig. 3.34 where the broken lines represent the raw experimental data. The solid lines are the difference between the data (after four-fold averaging to take advantage of the crystal symmetry) and the minimum intensity in each angular distribution. Each azimuthal distribution is drawn for a different polar angle of emission. This angle is indicated on each plot and is measured from the plane of the crystal to the measurement direction. The four-fold symmetry of the emission pattern is quite clear and it reflects the symmetry of the O atoms with respect to the underlying crystal structure. Surface structure analysis using this method is an important complementary method to LEED because of the use of the high X-ray brightness of synchrotron storage rings to reduce the data acquisition time for a result such as is shown in Fig. 3.34.

Surface extended X-ray absorption fine structure (SEXAFS)

If a beam of X-rays is transmitted through a solid and their absorption coefficient is measured it is found that a plot of absorption versus incident X-

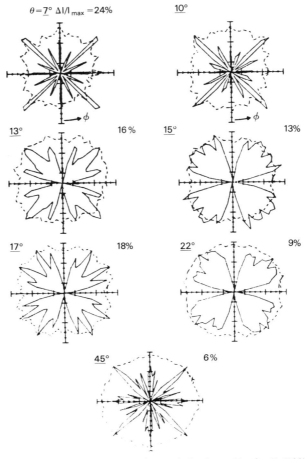

Fig. 3.34 Azimuthal distributions of O 1s photoemission intensities for Cu(100)c2-O for seven polar angles. (By courtesy of C. S. Fadley.)

ray energy exhibits considerable fine structure. Most obvious are the absorption edges as the X-rays rise to a sufficient energy to excite a new core level in the solid. Above each edge there are oscillations in the size of the absorption coefficient extending for 100–200 eV above the edge. This extended X-ray absorption fine structure (EXAFS) arises because photoelectrons excited by X-rays are backscattered from surrounding atoms and interfere with the outgoing photoelectron wave. The interference modifies the photoionization cross-section and hence the X-ray absorption coefficients. As the emitted photoelectrons vary in energy with increasing X-ray energy the interference conditions change correspondingly and the X-ray absorption coefficient is modulated. The Fourier transform of this absorption coefficient contains peaks separated by the interatomic distances. This EXAFS measurement thus gives information about the arrangement and spacings of atoms around the absorbing atom. It is particularly useful because it does not require a single-crystal sample, but rather that there be a site preference or particular local environment around the absorbing atomic species.

The EXAFS method is not sufficiently surface-specific for surface crystallography, but it can be adapted to observe the EXAFS oscillations indirectly so as to increase the surface sensitivity. If the X-rays create core holes, as they must, then not only does the X-ray absorption coefficient vary, but also the Auger electron yield, and the secondary electron yield must vary in a related way. Because of the small escape depths of these electrons the method becomes surface EXAFS or SEXAFS. Measurements of Auger or secondary electron yield, as a function of incident photon energy, give information about the geometry of a surface atomic arrangement, and an example is shown in Fig. 3.35.

Of course, the Auger yields are small and a high-brightness, variable-energy photon source is required to obtain an adequate signal-to-noise ratio and explore the SEXAFS oscillations. Such a source is provided by synchrotron radiation together with a monochromator (p. 86).

The polarization of the photon beam generated by an electron synchrotron source can be exploited to obtain more information about the structure of a surface with an ordered adsorbate. The example shown in Fig. 3.36 due to Warburton *et al.* (1987) is some of the data for Ni(110)c(2 × 2)S. They collected grazing incidence data with the **E** vector in the [001] azimuth and near normal incidence data with the **E** vector in the [001] and the [110] azimuths. The data are best described by the S atom being situated in a site which is a rectangular hollow in the surface having four Ni nearest neighbour atoms in the surface layer and a single Ni atom beneath in the second layer of Ni atoms. The polarization dependence of the backscattering amplitude allows the determination of the distance between the S atom and the Ni atom beneath it. The result of the theory–experiment comparison is not only that the S atoms occupy this rectangular hollow and have a coordination number of 5, but also that the S atoms are 2.23 ± 0.04 A from the Ni atoms in the surface plane and 2.31 ± 0.02 A from the Ni atoms beneath them.

Predicting the surface crystallography

In order to understand the wealth of results about surface crystallography which have been obtained using all the techniques outlined above it is important to have some theoretical treatment of just which surface relaxations and reconstructions are to be expected for any particular single crystal surface of a specified substance. On the one hand, the theory provides suggestions for possible surface structures which can be tested by experiment. On the other, any discrepancies between the theoretical predictions and observations can be

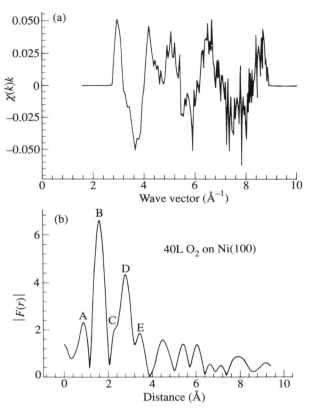

Fig. 3.35   (a) The EXAFS oscillations above the O 1s edge for 40L of O adsorbed upon Ni(01). The vertical scale is proportional to the X-ray absorption, which was measured by detecting the secondary electron yield at 3 eV. As X-ray absorption rises so does the secondary electron yield. The horizontal scale is proportional to the wave vector of the incident photons. (b) The Fourier transform of (a). The peaks occur at interatomic spacings in the surface region. B corresponds to the 1.9 Å spacing of O and its Ni nearest neighbours. (By couresty of J. Stohr 1979.)

instructive in helping to carry out a more precise or reliable experiment or can suggest new ways of approaching the theory.

One approach to a theory for the surface structure is to set up a calculation of the total energy of the semi-infinite solid. This energy is then minimized by moving the surface atoms about relative to the underlying (bulk) structure, which is not changed. Each set of coordinates for the surface atoms requires a separate calculation of the total energy of the system. A configuration of atoms is sought by repeating the calculation for many different geometries until a minimum is found in the total energy. Since the changes in total energy due to movements of surface atoms can be small compared to the total energy itself the calculations must necessarily be carried out with high precision. This means, in turn, that the physics of the whole problem has to be properly

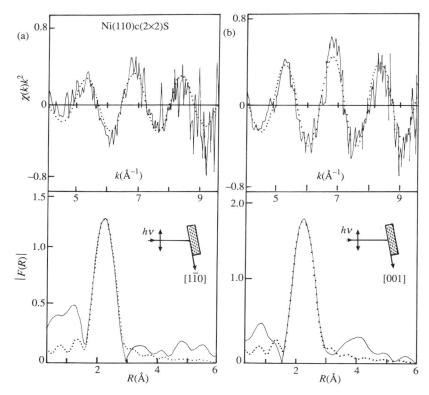

Fig. 3.36 KLL Auger yield SEXAFS spectra of Ni(110)c(2 × 2)S obtained at 10° off normal incidence of the photon beam with the **E** vector in the (a) [110] azimuth and (b) [001] azimuth. This Auger peak occurs near 2100 eV. In the upper figures the experimental SEXAFS oscillations in $\chi(k)$ weighted by k^2 (solid line) are compared with the theoretical best fit (dotted line). The corresponding lower section shows the modulus of the Fourier transform. (By courtesy of Dr G. Thornton.)

included. Very large supercomputer programs have been used to carry out such calculations.

Chadi (1978) has developed a theory for the surface structures of covalent and ionic materials. It focuses upon changes in the total energy resulting from atomic displacements and has given predictions of semiconductor surface structures which agree well with those determined using LEED. He expresses the total energy E_t of a system of electrons and ions as

$$E_t = E_{ee} + E_{ei} + E_{ii}, \tag{3.18}$$

where E_{ee}, E_{ei}, and E_{ii} are the energies due to electron–electron, electron–ion, and ion–ion interactions, respectively. Because existing calculations of the electronic band structure must take into account the electron–ion and electron–electron energies, it is useful to recast eqn (3.18) using a band structure energy E_{bs}, which is defined as the sum over the energies of all occupied states in the solid. This gives

$$E_{bs} = E_{ei} + 2E_{ee},\tag{3.19}$$

the factor of two arising because each electron–electron interaction is counted twice in summing the energies of all occupied states. If this is substituted into eqn (3.18) then the total energy can be expressed as

$$E_t = E_{bs} + U, \quad \text{with} \quad U = E_{ii} - E_{ee}\tag{3.20}$$

The advantage of this last equation is that, provided the ions are separated by more than the Thomas–Fermi screening distance, the system (ion + electrons) is nearly neutral and U is close to zero. Therefore approximations can be used to estimate the change to E_t due to movements of the atoms by holding E_{bs} constant (the band structure energy of the bulk material) and having a simple model to see how U changes. Indeed, Chadi was able to account for the observed atomic displacements with only a two-parameter model describing U. This is not a self-consistent technique for the calculation of the surface structure (see Chapter 4) but it does suffice to explain many aspects of the prediction of the crystallography without going to the complexity of full self-consistent calculations of the energy of the system.

This method was applied to a study of the first three layers of the (110) surfaces of Si, Ge, GaAs, InP, ZnSe, ZnTe, and InSb. The changes in the calculated total energies from the bulk termination of the structure to the lowest energy found were as large as -0.51 eV per surface atom for GaAs(110). The minimum total energy structure was found to have the As atoms out towards the vacuum and the Ga atoms in towards the solid. These displacements occur in such a way that the Ga–As bond length stays approximately constant but the bond rotates out of the surface plane by about 27°. Roughly the same rotation is calculated for all the materials listed above. The structure is sketched in Fig. 3.37. This prediction is in good agreement with experimental determinations of the surface structures of these materials.

Some selected surface structures

A huge number of surface structures have been determined. Over 200 have been determined by LEED crystallography alone. For detailed evaluation of all this work and for references the reader would be wise to consult the book by Van Hove *et al.* (1986) and the thorough compilation by Watson (1987). Here, a few results have been abstracted from the tables in those publications to draw out some of the more striking observations.

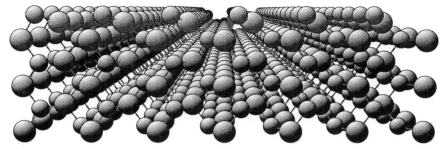

Fig. 3.37 A three-dimensional diagram of the bond rotations occurring in a simple one layer model of the surface of GaAs(110). The As atoms are represented by the smaller spheres. They can be seen to have moved out towards the vacuum (at the top of the figure) in the top atomic layer. (By courtesy of Dr S. P. Tear.)

Clean metallic elements

Three low-index surfaces of the clean f.c.c. metals have structures summarized in Table 3.1. The first striking observation is that the (100) and (111) faces are very close to being simple bulk terminations of the crystal. On the other hand, the more open (110) usually shows an inward relaxation and even a reconstruction for the heavier metals Pt and Ir. Indeed, careful surface

Table 3.1 Some clean f.c.c. metals. These data were abstracted from the compilation by Van Hove *et al.* (1986). The references to the original work are contained therein. Pt(110) reconstructs with a clean (2×1) surface which is thought to be due to a missing row of surface atoms. Ir(100) reconstructs with a clean (5×1) surface thought to consist of a buckled quasi-hexagonal layer of surface atoms

Element	(100)	(110)	(111)
Al	0	-12.5 ± 2.5	0
Ag	0	-6	0
Au	$(1 \times 1)0$	-15	0
Cu	-0.5 ± 1	-10 ± 3	0
Ni	0	-1.5 ± 5	0
Pt	(1×1) 0.5 ± 0.5	(2×1)	0.3 ± 1.5
Ir	(1×5) Quasi-hex.	(2×1) -15	-2.5

crystallography by Andersen *et al.* (1984) shows that Ni(110) and Al(110) have oscillatory relaxation at their surfaces. The surface layer is contracted towards the solid, the next layer is displaced by a smaller amount towards the vacuum, the next contracted by an even smaller amount towards the solid, and so on.

The h.c.p. and b.c.c. metallic elements also tend to have very small relaxations at their low-index surfaces and oscillatory relaxation has not yet been observed in any of these materials.

Clean semiconductors

As seen in Fig. 3.15, the (111) surface of silicon reconstructs to form a 7×7 surface mesh. Another example is shown for CdTe (110) in Fig. 3.20, where the bond between surface Cd and Te atoms is shown to rotate out of the mean surface plane. These rather dramatic reconstructions are typical of the surfaces of semiconductors. They come about because the bonding is dominantly covalent and so is very directional. At a free surface and in the absence of any reconstruction there may be electrons spending time outside the solid in bonds dangling into the vacuum. The electrostatic effects of such dangling bonds are very considerable and give rise to an increase of the surface free energy over what it might be should the atoms move and the bonds rearrange.

Table 3.2 outlines some of the angles and distances involved in the reconstructed surfaces of the zinc blende family of structures which cleave on their (110) planes and for which GaAs(110) is an archetype. These surfaces all show a bond rotation of the type shown in Fig. 3.37, which varies in size for different materials.

Table 3.2 The cleavage faces of the zinc blende structure semiconducting compounds

Surface	Orientation	Lattice constant (nm)	Bond rotation (°)
AlP	(110)	0.55	25.2
AlAs	(110)	0.57	27.3
CdTe	(110)	0.65	30.5
GaAs	(110)	0.57	27.3
GaP	(110)	0.55	27.5
GaSb	(110)	0.61	30
InAs	(110)	0.61	31
InP	(110)	0.59	28.1
InSb	(110)	0.65	28.8
ZnS	(110)	0.54	25
ZnTe	(110)	0.61	28

Clean insulators

At first sight it might be thought impossible to measure LEED intensities from the surface of an insulating material, because the surface may charge up electrostatically as the low-energy electron beam strikes it. However, the secondary electron emission coefficient of some insulating materials is so high that more electrons leave the surface into the vacuum than arrive in the incident beam. Thus the surface can be positively charged. The electric field around the sample then causes some of the lower energy electrons to be attracted back to the surface. The whole system can come to equilibrium with the surface at a small positive potential with respect to ground and a negative space charge cloud in front of the surface. This positive potential accelerates the electrons into the solid and so shifts the $I(V)$ curves towards lower energies, but they are still observable. The divalent metal oxides (MgO, CaO, NiO, MnO) all have high secondary electron yields—indeed, MgO(100) can have a yield as high as 24 secondary electrons per primary electron. For this reason their surface crystallography can be established by LEED.

The surfaces of these materials might be expected to rumple by outward relaxation of the large, easily polarizable, O^{2-} ion and inward relaxation of the small almost unpolarizable metal ion. The square mesh of the bulk exposed plane is not expected to change. Table 3.3 summarizes some structures for the divalent metal oxides. Very small rumpling effects have been observed for MgO(100), but it is a good approximation to regard the cube faces of these materials as unrelaxed and unreconstructed in spite of the huge polarizability difference between the two kinds of ion present. Little theoretical work on this problem has been reported.

Table 3.3 The cube faces of some divalent metal oxides

Surface	Relaxation (%)	Rumple (%)
CaO(100)	-1 to -3	0 to $+2$
CoO(100)		
MgO(100)	0 to -3	0 to $+5$
NiO(100)	0 to -3	0 to -3

Adsorbates

By far the most surface crystallography has been carried out on the structure of a wide variety of atoms or molecules adsorbed onto the low index crystal faces of many compounds and elements. A tabulation of many of the results of this work up to 1987 has been published by Watson (1987) and is also discussed by van Hove (1991). The subject of adsorption is introduced in Chapter 6. A few examples of the ordered adsorbates that can form in the case of oxygen on two low-index faces of some metals are shown in Table 3.4. The adsorbate atoms

Table 3.4 Selected adsorbate systems for which oxygen has been adsorbed upon a metallic substrate. The number 3 under 'Site' indicates adsorption into a three-fold coordination with an atom of the substrate directly below the adsorbate atom. The number 4 means a four-fold coordination. The number 2 means a two-fold site (i.e. a bridge site). d_0 is the spacing of the overlayer of oxygen atoms from the first layer of metal atoms in the direction normal to the surface. δd is the percentage difference of from the bulk spacing of the first to the second metal layers. The data have been abstracted from the extensive compilation of Watson (1987)

Material	Mesh	Site	d_0 (Å)	δd (%)
Al(111)	(1 × 1)	3	1.46	0
Rh(111)	(2 × 2)	3	1.23	0
Co(100)	c(2 × 2)	4	0.80	0.0
Cu(100)	c(2 × 2)	2	1.4	—
Ni(100)	c(2 × 2)	4	0.90	0
Ta(100)	(1 × 3)	—	2.0	−5
W(100)	Disordered	4	0.55	0

are often found to be situated in the high-symmetry sites of the metallic surface, presumably because these provide the greatest number of bonds to the substrate atoms. Sometimes the adsorbed atoms remove the surface relaxation exhibited by the clean metal and most of the cases in Table 3.4 are of this kind. Sometimes the presence of the adsorbate induces reconstruction in the metal surface—this appears to be the case, for instance, with Fe(110)p(2 × 2)–S and Ni(100) (2 × 2)–C. Another possibility which is sometimes observed is that the adsorbate atoms penetrate the metallic surface and form an underlayer beneath the surface layer of metal atoms. Examples of this phenomenon are provided by Ti(0001) (1 × 1)–N and Zr(0001) (2 × 2)–O.

Summary: surface structure and composition

The techniques described in Chapters 2 and 3 can be used to characterize the structural and chemical state of a surface. Electron spectroscopy provides a powerful technique for establishing what species are present at the surface and in what quantities. AES is particularly sensitive if the initial state ionizations are created by an electron beam. Sensitivities of the order of 1 per cent of an atomic monolayer are possible. Quantitative analysis is more difficult and is confined mostly to systems where the surface species are in the form of uniform layers upon the substrate. SIMS is much more sensitive than AES, but quantification can be more difficult (and less precise) because of very large matrix effects modifying the ion yields. Information about the type of chemical bond present can be derived fron studies of the positions and profiles of the features in the characteristic electron spectra. XPS, AES, and SIMS can all be used to image surfaces in such a way that the contrast is chemically specific.

The size and symmetry of the surface unit mesh of a crystallographically ordered surface can be determined with an accuracy of about 1 in 10^3 by LEED, RHEED, or grazing incidence X-ray diffraction. SEXAFS yields quantitative estimates of the nearest neighbour coordination and distances. Specification of the positions of each type of atom within the unit mesh is more difficult but a large number of surface structure determinations are now available. The imaging of individual atomic positions in the surface is possible with STM, which forms an important ancillary tool to the surface diffraction techniques.

4
Surface properties: electronic

The earlier chapters have concentrated upon the techniques required to obtain information about what kind of atoms are present at a surface and where they are situated. As the need for other techniques arises they will be discussed, but now attention will be turned towards the ways in which the properties of surfaces can be related to their composition and structure.

Some theoretical considerations

The calculation of the electronic states at a surface is bound to be more complex than the corresponding calculation for electrons inside a solid. The latter is complicated enough, because the least tightly bound electrons are free to move in the periodic potential established by the arrangement of nuclei and tightly bound electrons—the ion cores. Furthermore, the moving electrons repel each other because of their negative charges. Thus, Schrödinger's equation has to be solved for a many-body periodic potential. This is rarely analytically possible. Various approximations are sought whose complexity depends upon the physical property being described and the precision needed to compare this property with experimental measurements. Free electron theory treats the periodic arrangement of positively charged ion cores in a crystal as a smeared out, generally attractive, potential for the gas of free electrons. Nearly free electron theory improves this model by having a weak periodic potential superimposed upon the uniform potential in the well formed by the piece of solid material. Better approximations are provided with pseudopotential approaches or tight binding approximations. Descriptions of these models for the electronic states in a three-dimensionally periodic arrangement of atoms can be found in many textbooks of solid state physics (e.g. Rosenberg 1974; Kittel 1986).

The free electron theory of solids treats a fictitious material often referred to as *jellium*. In order to describe the surface the simple picture is that the uniform positive potential in the solid is cut off abruptly at the surface. This problem was solved self-consistently by Lang and Kohn in 1970 and their calculated variation of the electron density in a free electron metal with three outer electrons per atom (like aluminium) is shown in Fig. 4.1(a). Because the electrons are free to move inside the solid they attempt to screen out the attractive potential as it turns on at the solid surface. This results in an enhancement of the electron density near to the surface and an oscillation in

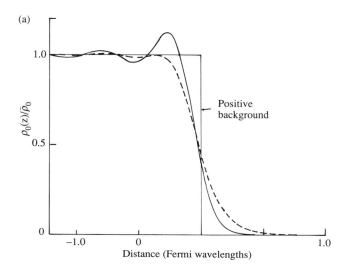

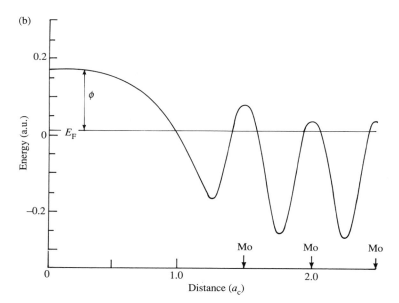

Fig. 4.1 (a) The variation of the charge density of electrons near to the surface of an abruptly terminated uniform potential in a free electron solid. Unlike the calculations due to Chadi outlined in Chapter 3, this work is a self-consistent model of the charge distribution and the potential at the solid surface. (After Lang and Kohn 1970.) (b) The variation of the electrostatic potential normal to the surface of Mo(100) taking into account the periodic potential of the Mo ion cores. (After Kerker *et al.* 1978.)

density which dies away rapidly into the solid. This oscillation is more pronounced for materials with lower electron densities. The spreading of the electron wave functions out into the vacuum, which can be seen in Fig. 4.1(a), leads to a surface dipole moment which contributes to the work function of the surface (see below).

The free electron approximation is useful to begin to understand the properties of electrons both inside and at the surface of a solid. However, real materials have periodic atomic arrangements and so the electrons move in a periodic electrostatic potential. It is this periodicity which gives rise to the band gap in the electronic structure and so to electron diffraction (Chapter 3) and, of course, to the existence of metals, semiconductors, and insulators. The theory of the band structure in a periodic potential which terminates at a surface can become very complex because of the large number of electrons which interact with each other via Coulomb and exchange forces. The subject is introduced in Kittel (1986) for three-dimensional potentials and by Inglesfield (1982) for the situation at a surface.

The simplest metals are those that are s–p bonded and have rather weak periodic potentials. Examples of such materials are aluminium and the alkali metals. Wang *et al.* (1981*a*) have presented results for the Al(111) surface, and the contours of charge density that they find are reproduced in Fig. 4.2.

The striking prediction here is that the electron charge density is rather smeared out near to the surface (note the lack of very much ripple in the contours of Fig. 4.2 until one reaches positions quite close to the Al ion cores). This smoothing of the surface charge density is the result of the opposing effects of the potential energy drawing the electrons towards the ion cores and the kinetic energy of the electrons, which is reduced when the electrons are spread out. The work function ϕ (see below) is one experimentally measurable quantity which emerges from the kind of calculation needed to produce contour maps like Fig. 4.2. Provided that an adequate approximation is included in the theory to account for the correlation effects of the Coulomb and exchange forces between the electrons, then the theoretical and experimental values of ϕ can agree to precisions of about 0.1 eV.

Another prediction from such theories is that there can be a situation in which charge can accumulate at a surface in positions between atomic planes. This charge can reduce the energy of the surface because the s-like states which concentrate at the ion cores have higher energy than the p-like states of electrons collecting between the atoms. This is a description of a *surface state*. It is a solution of Schrödinger's equation in the region near to the surface where the potential varies periodically in only two dimensions. For the simple metals these surface states show up in experiments measuring the angular distribution of photoelectrons emitted from a single-crystal surface (see below). There are two simple views used to classify surface states. At one extreme there is a *Shockley state*. Here, solutions to Schrödinger's equation exists in which there are waves with energies in a forbidden band gap whose

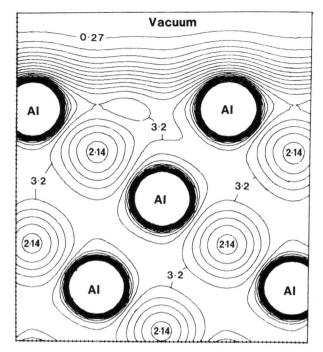

Fig. 4.2 The calculated contours of the electron charge density near to an Al(111) surface. (After Wang *et al.* 1981*a*.)

amplitudes decay away exponentially into the crystal. They have complex wave vectors. The surface states seen in calculations for the simple metals often show such states and the case of Al(111) seen in Fig. 4.2 is of this kind.

If the Fourier coefficients of the potential in a crystal are larger than those for simple metals then the band gaps become larger, and even nearly free electron theory becomes completely inappropriate. The electrons are best regarded as being tightly bound to the ion cores. This certainly occurs in insulators and is even a reasonable approximation for many semiconductors. When tight binding of electrons is occurring then solutions to Schrödinger's equation at a surface can include eigenfunctions whose energy corresponds to a band gap but which are localized on the surface atoms. This extreme is referred to as a *Tamm state*. It is the kind of surface state one might expect to find at a semiconductor surface where there are broken bonds.

In a real material the situation is at neither of the free electron or tight binding extremes and the actual surface states found are neither simple Tamm states nor simple Shockley states. Some examples of the results of electronic state calculations are shown in Fig. 4.3. Figure 4.3(a) shows a charge density contour map for Ni(100)—a transition metal surface, and for Si(001) 2 × 1—a

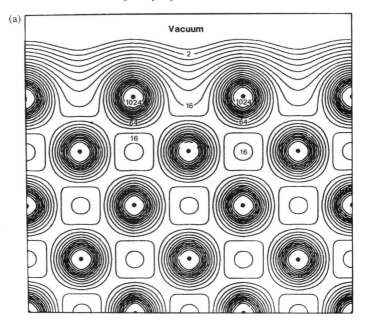

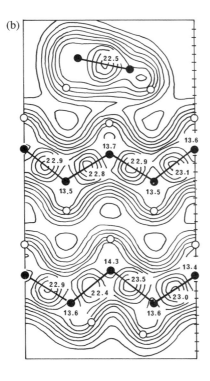

Fig. 4.3 (a) The charge density contours for Ni(001) calculated using a self-consistent slab method. (By courtesy of F. J. Arlinghaus *et al.* 1980.) (b) A similar plot for Si(001) reconstructed to the 2 × 1 surface structure. (By courtesy of J. Ihm *et al.* 1980.)

semiconducting surface. The progression of the localization of the charge density from simple metal to transition metal to semiconducting is clear in the comparison between Fig. 4.2 and Fig. 4.3.

Contact potential and work function

One macroscopic electronic property of a surface which can be measured with some precision is the work function, ϕ. It is convenient to define ϕ for a metal in terms of the free electron model of its electronic behaviour (e.g. Rosenberg 1974). It is the difference in energy between an electron at rest in the vacuum just outside a metal and an electron at the Fermi energy. In other types of materials, such as semiconductors and insulators, it can be regarded as the difference in energy between an electron at rest in the vacuum just outside the solid and the most loosely bound electrons inside the solid. It is clearly an important parameter in situations in which electrons are removed from a solid. For example, it affects thermionic emission, the contact potential between solids, emission of electrons in high electric fields, and the bonding of impurity atoms to a surface. Further, it provides an important test of the success, or otherwise, of an electronic structure calculation.

There are several physical factors contributing to ϕ (see e.g. Riviere 1969). In the first place it can be seen, using Fig. 4.4, that the value of ϕ depends upon the depth W of the attractive potential for the conduction electrons inside the solid. This is a bulk property determined by the attraction for its electrons of the lattice of positive ions as a whole. This contribution to W will depend upon the type and arrangement of positive ions in the lattice. It is an energy of the order of a few electron volts.

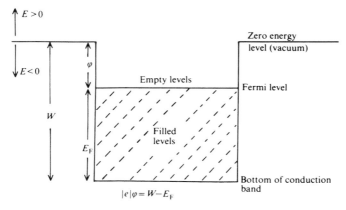

Fig. 4.4   The distribution of conduction electron energies in the free-electron model of a metal. E_F is the Fermi energy; ϕ is the work function; W is the potential well bonding the conduction band electrons into the solid.

In addition, there are specifically surface contributions to W. This implies that it is wrong to draw W as independent of position in the solid, as in Fig. 4.4, but that W should change in the region of the surface. One such contribution is the *image potential*. Electrostatic theory shows that a charge $-e$ outside a conductor is attracted by an image charge $+e$ placed at the position of the optical image of $-e$ in the conducting plane. If $-e$ is at a distance z from the plane, the image force is thus $e^2/4\pi\varepsilon_0 (2z)^2$. This force is experienced by an electron escaping into the vacuum, and so is a contribution to the work function. It can be shown to be negligible beyond 10^{-6}–10^{-5} cm away from the surface. This classical description of the image force breaks down when the escaping electron is very close to the surface (<0.1 nm away) and a quantum mechanical description is necessary for the interaction of this electron with those remaining in the surface (e.g. Gadzuk 1972).

A second surface contribution to W is the strength of a possible surface double layer. As the surface atoms are unbalanced, because they have matter on one side and not on the other, the electron distribution around them will be asymmetrical with respect to the positive ion cores. This leads to a double layer of charge, as indicated in Fig. 4.5. Two important effects of this double layer are that it results in the work function being sensitive to both surface contamination and the crystallographic face exposed. The contamination will affect ϕ because it will modify the double layer in a way depending upon the affinity of the contaminant for electrons. As the electron affinity of atoms depends upon their type, so ϕ will vary according to the type of contaminant. Also, the orientation of the exposed crystal face will effect ϕ because the strength of the electric double layer depends upon the density of positive ion cores, which in turn varies from one face to another. The contribution of the double layer to ϕ is of the order of a few tenths of an electron volt.

These arguments lead to the expectation that the work function ϕ will depend upon which crystal face is being studied and the extent to which it is atomically clean.

The *contact potential* between two metals (e.g. Rosenberg 1974) is given simply by the difference between the work functions of the two metals.

The measurement of work functions

The absolute measurement of work function is difficult. The thermionic emission of electrons depends exponentially upon ϕ through the Richardson–Dushman equation (e.g. Rosenberg 1974), and an experiment to measure thermionically emitted current as a function of temperature can be used to derive a value for ϕ. Unfortunately, measurable thermionic emission occurs only at high temperatures (ϕ is usually greater than 2 eV), and use of this technique is thus confined to those materials that have low vapour pressures at high temperatures (e.g. Ni, Mo, Pt, W, Ta). Another absolute measurement can be obtained from observations of the photoelectric yield (the total emitted

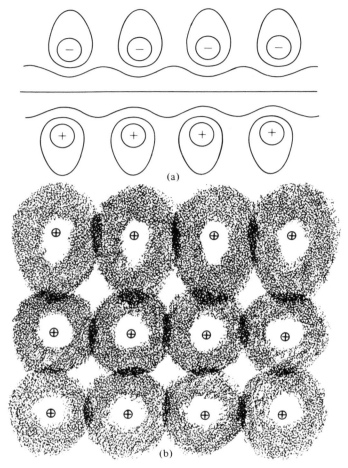

Fig. 4.5 (a) Equipotential lines in a general electric double layer. Should the centre of this layer be a surface plane there would be a potential step on passing across it. (b) A possible distribution of charge in a simple cubic lattice of positive ions. The distortion of the electron cloud at the surface causes an increased electron density between the cores compared to the bulk. A double layer with either net positive or net negative charge at the outside may be created by such a distortion.

photoelectron current) from the material as the frequency v of the incident radiation is varied. It can be seen from Fig. 4.4 that, at the absolute zero of temperature, no photoemission should occur for $hv < \phi$ and that, above the threshold v_0 given by

$$hv_0 = \phi, \qquad (4.1)$$

the photoelectric yield should rise sharply as electrons at the top of the conduction band are excited into the vacuum. Fowler (1933) was the first to show how such a technique could be used to obtain accurate values for ϕ. Most clean metals have work functions of the order of a few eV, and so the incident radiation for determination of the photoelectric threshold is in the ultraviolet region of the spectrum.

If a large electric field is applied to a cathode at room temperature it is possible to draw a current of electrons from it. The electrons from the metal escape to the vacuum through a thin potential barrier by quantum mechanical tunnelling (Fig. 4.6). The size of this field-emission current depends upon the work function of the cathode. The field-emission microscope is based upon this phenomenon, and its use is described later in this chapter.

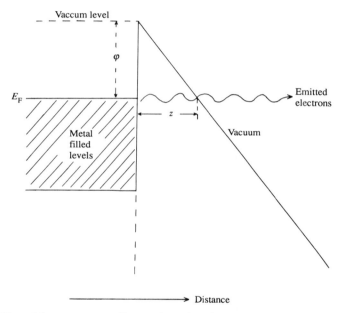

Fig. 4.6 Potential energy versus distance for a free-electron metal with its surface as one electrode in a strong electric field E. For sufficiently large E the potential barrier thickness z can be so low that electrons have high probabilities of tunnelling from near the Fermi level E_F into the vacuum.

Two techniques which are useful for determinations of the *relative* work function are retarding potential difference (RPD) and contact potential difference (CPD) methods. A sensitive vibrating probe CPD technique is explained in Fig. 4.7. If the probe is made of a material with known work function (commonly tungsten or platinum) then the technique can be used to determine the CPD V_{12} between the probe and the sample and hence ϕ for the sample. Even if the work function of the probe is not known, the technique is useful in measuring the change in ϕ of the sample as its surface is changed during an experiment. Because the technique is used to detect the null point when $V_B = -V_{12}$, it can be developed, with electronic techniques, to be extremely sensitive. With care, changes in ϕ of only 1 meV are detectable. In addition, it is convenient to use in conjunction with electron energy analysers and LEED optics, as described in Chapters 2 and 3.

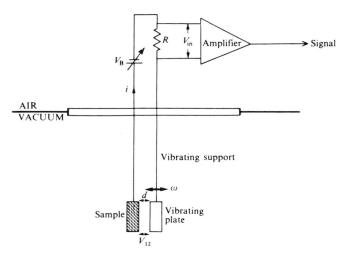

Fig. 4.7 The vibrating probe technique for determination of relative work functions using the contact potential difference (CPD). V_{12}, contact potential; C, capacitance; A, plate area; q, charge; d, plate separation; d_1, amplitude of vibration; ω, frequency of vibration.

$$d = d_0 + d_1 \sin \omega t,$$
$$q = C(V_B + V_{12}) = (A/\varepsilon_0 d)(V_B + V_{12}),$$
$$V_{in} = iR = (\mathrm{d}q/\mathrm{d}t)R = -(AR/\varepsilon_0 d^2)(V_B + V_{12})d_1 \cos \omega t.$$

$$\boxed{V_{min} = 0 \text{ if } V_B = -V_{12}}$$

For high sensitivity use low d, high R, and high ω.

Work function changes can be measured in the LEED apparatus of Fig. 3.14 using an RPD method. The sample surface is bombarded with electrons, and the total current i_c backscattered to the screen S is measured as a function of the retarding potential $-V$ applied to the grid G2. The total potential difference acting on the electrons is $-V + V_{12}$ where V_{12} is the contact potential between the sample and the screen. If the sample surface is now changed in some way and has a new value of work function then the curve of i_c versus V will be displaced by the change $\Delta\phi$. Again, this method is convenient because it is easily combined with other experiments.

Some examples of experiments involving work functions

Dependence of ϕ upon crystal face   Because of its suitability as an emitter in
electronic guns, tungsten has been the subject of careful study. The values of ϕ
for different low-index faces of tungsten are shown in Table 4.1. The smallest
values of ϕ are associated with least densely packed face. Presumably, the
rearrangement of electron density on the least densely packed face is such as to

Table 4.1   Work functions of
tungsten

Crystal face	Work function (eV)
(111)	4.39
(100)	4.56
(110)	4.68
(112)	4.69

produce a net positive charge on the vacuum side of the surface and so a net
transfer of electrons to the inside and thus a lowering of the work function.

   This variation of work function from one crystal face to another is
demonstrated elegantly in the *field-emission microscope* (FEM). This instrument is identical in construction to the FIM described on p. 85. Now,
however, no gas is admitted to the vacuum chamber and the sign of the
potential on the sample tip is arranged so that electrons are accelerated out of
it by a very high local electric field (typically 4×10^7 V cm^{-1}). The theory of
this process is described by Muller (1970). The current leaving the region of the
tip surface where the work function is ϕ is roughly proportional to
$\exp(-A\phi^{3/2})$, and so is a very fast function of ϕ. The brightness observed on
the fluorescent screen is thus a function of the value of ϕ at that place on the
tip. Since, in practice, a sharp tip will be faceted so as to expose small flat areas
of different crystal faces, the FEM image will consist of patches corresponding
in position to these faces and of brightness depending upon the work function
of each face. An FEM image of a tungsten tip is shown in Fig. 4.8, which is
labelled with the Miller indices of each facet giving a bright spot. The brightest
spots correspond to the lowest values of ϕ in Table 4.1.

   The changes in ϕ produced by adsorbed atoms or molecules can be seen in
the FEM and the diffusion of these adsorbates over the tip surface followed.
Spatial resolutions of about 2.5 nm are possible.

Dependence of ϕ upon contamination   If the adsorption of atoms or molecules
results in the transfer of charge to or from a surface then the work function will

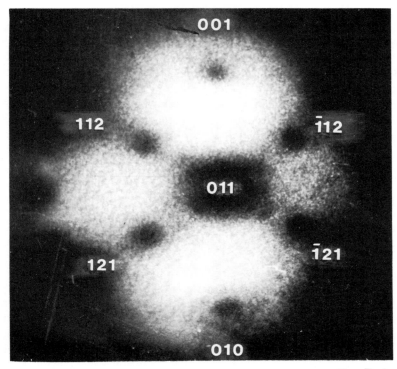

Fig. 4.8 An FEM image of a (011) single-crystal tungsten tip of radius 210 nm. The patches in the image are labelled with the Miller indices of the crystal facets to which they correspond. (By courtesy of Dr H. Montagu-Pollock, University of Lancaster.)

change. The adsorbed species may be polarized by the attractive interaction with the solid surface, or, more extremely, it may be ionized. If it is polarized with the negative pole towards the vacuum the consequent electric fields will cause an increase in work function. This can be seen by considering the effects of a deeper positive potential at the edge of the well in Fig. 4.4. Conversely, if the positive pole of the polarized adsorbate is outwards then the work function of the substrate will decrease. Similarly, an adsorbate which is in the form of positive ions will have transferred electrons to the substrate and so decreased its work function. So long as the number of adsorbed ions (or polarized atoms or molecules) is low enough for interactions between them to be negligible, the change in work function will be proportional to the number of ions adsorbed. The variation of work function with type of adsorbed species is discussed by Somorjai (1972).

The effects of an adsorbate upon work function are clearly demonstrated in the work of Adams and Germer (1971), who studied the adsorption of molecular nitrogen upon W(100), (310), and (210) surfaces. By correlating work function and LEED observations with measurements of the amounts of nitrogen desorbed into a mass spectrometer (flash desorption) after each experiment they were able to obtain the data shown in Fig. 4.9. By studying the variation of the LEED patterns with the dose (rate of arrival of molecules × time of exposure) they were able to conclude that adsorption ceased after half a monoloayer of nitrogen had arrived. This *saturation coverage* was removed by heating the crystal quickly to desorb the nitrogen. Some of the nitrogen is collected by a mass spectrometer the output of which is then calibrated in terms of the known saturation coverage. Thus the data for the abscissa of Fig. 4.9 can be obtained.

Surface states and band bending

As described earlier in this chapter, the three-dimensional periodicity of the potential is lost at the surface, and travelling wave solutions of the Schrödinger equation can be found which are situated within the band gap of the bulk solid. These special solutions are waves which can travel parallel to the surface but not into the solid. Thus they are localized at the surface and can have energies within the band gap of the bulk band structure. These solutions are known as *surface states*. If charge (either electrons or holes) is situated in these surface states it will result in electrostatic fields penetrating into the solid, and thus a varying electrostatic potential on passing from the surface to the interior. This varying potential distorts the band structure of the bulk solid—an effect which is called *band bending*.

Surface states can be associated not only with the termination of a three-dimensional potential at a perfect clean bulk exposed plane but also with changes in the potential due to relaxation, reconstruction, structural imperfections (such as emerging dislocations), or adsorbed impurities. If the charge associated with any of these surface states is different from the bulk charge distribution, then band bending will result. The surface states will affect the electrical properties of a surface by acting as a source (or a sink) of electrons and the chemical reactivity by modifying the affinity of the surface for electrons.

Quantitative estimates of the effects upon band shapes of charge localized in surface states are quite difficult to make. If the free-carrier density in the material is very high then compensation of the charge at the surface by flow of carriers out of the bulk is effective (the conduction band electrons 'screen' the surface charge). Thus, metals tend to have negligible band bending. In semiconductors and insulators such complete compensation is not possible because of their much lower free-carrier densities, and the existence of occupied surface states can result in surface potentials of the order of millivolts or volts, the band bending effects extending some micrometres into the solid. The simplest theories for surface states are outlined, for instance, in Wert and

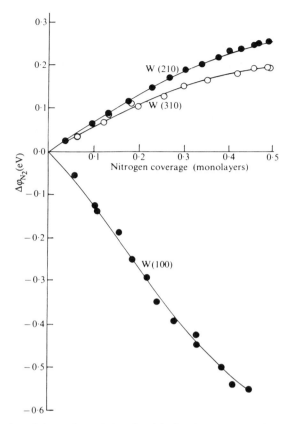

Fig. 4.9 Variation of change in work function $\Delta\phi$ of various tungsten single crystal faces with coverage of nitrogen. (After Adams and Germer 1971.)

Thomson (1970) and for band bending in semiconductors by McKelvey (1966).

If the potential varies on passing from the surface to the interior of a solid then the surface must have a capacitance. Measurement of the surface capacitance is one of the possible probes of the electronic properties of a surface. If a parallel plate capacitor (Fig. 4.10) is constructed such that its geometrical capacitance is small compared with the surface capacitance C_s then changes in C_s can be measured as a function of applied electric fields, coverage of various adsorbates, or exposure to electromagnetic radiation. The electronic properties of the surface can be inferred from the results of such experiments. Alternatively, the surface conductivity can be measured using contacts P and Q on the sample of Fig. 4.10 when charge is induced at the surface by applying a potential to the metal plate. If there were no surface

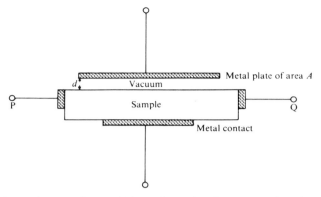

Fig. 4.10 A capacitor experiment which can be used to demonstrate the existence of surface states and the changes in charge trapped in them with varying surface conditions. Total capacitance C, surface capacitance C_s; geometrical capacitance C_g.

$$\frac{1}{C} = \frac{1}{C_s} + \frac{1}{C_g},$$

$$C_s = \frac{\varepsilon_o A}{\varepsilon_o A - Cd}\, C.$$

states the charge would be injected into the conduction band of the sample and the change in conductivity expected is predictable. In fact, only about one-tenth of this expected change is found, suggesting that relatively immobile charge is trapped in surface states.

Surface states can also be observed by the techniques of electron spectroscopy described in Chapter 2 and the CITS technique first outlined in Chapter 3. Semiconductor surfaces are particularly interesting in this respect because they show reconstruction.

An important example of a reconstructed surface is found for Si(111). The LEED pattern of a clean Si(111) surface (Fig. 3.15(b)) shows extra features at positions separated by one-seventh of the distance between the spots due to a bulk exposed plane, and thus atomic disturbances must be occurring in the surface with seven times the periodicity of the bulk atomic arrangement. The pattern is thus referred to as Si(111)7 × 7. A silicon atom in the bulk is covalently bonded to four nearest neighbours in a diamond cubic structure. At the (111) bulk exposed plane, each atom (Fig. 4.11) has one covalent bond 'dangling' into the vacuum in the direction of the surface normal. Such a dangling bond is energetically unfavourable, and the surface may be able to reduce its free energy by reconstructing in such a way as to reduce the number and/or energies in these bonds.

If a surface has dangling bonds or reconstructs so as to reduce the energy associated with the surface then the bonding for surface atoms will be different

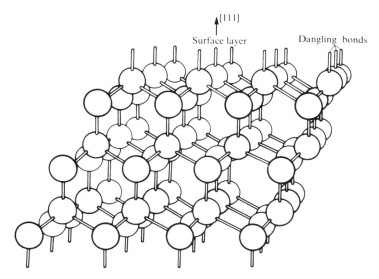

Fig. 4.11 'Dangling' bonds from the (111) surface of a covalently bonded diamond cubic structure.

from the bonding for bulk atoms, and surface states will exist. If electrons are energy-analysed (by one of the techniques in Chapter 2) after being scattered from the surface, some of them will have lost specific amounts of energy in exciting particular processes in the solid. These processes are mentioned on p. 22 and show up as features labelled as losses in Fig. 2.2(c)). The detailed study of such loss spectra is often referred to as *electron-loss spectroscopy*. As the primary energy is varied the losses move along the energy scale fixed by a particular energy difference from the elastic peak. The largest features in most loss spectra are the surface and bulk plasmon losses, but many other features can be observed which are due to excitation of an electron in the solid from its ground state to some empty state or band above the Fermi level. A careful study of the loss spectra of various Si(111) and Si(100) surfaces (Rowe and Ibach 1973) has shown that three features in the electron spectrum corresponding to losses of approximately 2 eV, 8 eV, and 14 eV can be associated with surface states. The correlation with surface states is made by noting the following: the way that oxygen adsorption decreases the size of the 14 eV feature, presumably because the oxygen atoms are using up the dangling bonds; the way the 8 eV peak is present only for well-ordered surfaces (as judged by LEED); and the way the strength of the loss features depends upon the surface roughness induced by Ar$^+$ ion bombardment. Thus, surface states, identified by peaks in loss spectra, can be correlated with particular bonds between silicon surface atoms.

Most information about surface states has been obtained by using angle-resolved photoelectron emission. This is a large and complex field of study, and the interested reader is referred to a review by Williams and McGovern (1984) for a more complete view.

Many features of a photoemission experiment can be understood by using a simplified approach known as the 'three step model'. Here, the excitation process of the photoelectron, its transport to the surface, and its escape through the surface to the vacuum are regarded as independent events which occur sequentially. Although this is an over-simplification of what actually happens in the excited, solid state environment of the experiment, it has served well to account for many properties of photoelectron spectra.

In the first step the probability of exciting an electron depends upon the electric vector potential of the incoming radiation as well as the wave functions of the initial and final states of the electron. Thus, both the angle of incidence and the polarization of the photons affect the excitation probability. Having been generated, a photoelectron may reach the surface without being scattered or may be scattered in inelastic or elastic collisions. The elastic scattering causes *photoelectron diffraction*, and will contribute to variations in photo-electron flux with emission angle. As mentioned in Chapter 3, this process requires LEED theory for its description. The inelastic scattering may be with small changes of energy (a few meV) caused by phonons (Chapter 5) or with more substantial changes caused by plasmon excitation or electron–hole pair generation. These loss processes result in a rather featureless background of secondary electrons below the peaks in a photoelectron spectrum.

If momentum conservation is applicable to the initial and final states of the photoelectron excitation then, in a reduced zone scheme for the one-electron states of the solid, only vertical or direct transitions are allowed (Fig. 4.12). An angularly resolved experiment allows the determination of the binding energy of the photoelectron (Chapter 2) through knowledge of the spectrometer energy calibration and the momentum of the photoelectron through knowledge of the orientation of the entrance slit of the spectrometer with respect to the sample's surface normal. If direct transitions are involved, the binding energy and momentum of the initial state can be deduced and so the dispersion, $E(k)$, of the initial state can be determined. For off-normal emission angles lack of information about the refraction effects at the surface make k inside the crystal uncertain, but this doubt is removed for normal emission. For this reason, photoemission along the surface normal is a very powerful method for determining $E(k)$ in that direction in the band structure.

The angularly resolved photoemission experiment is thus arranged with a source of monochromatic photons (ultraviolet resonance lamp, X-ray source, or synchrotron storage ring), a sample in a goniometer holder, and a dispersive energy analyser (Chapter 2) which can be moved to explore the emitted photoelectron distribution. The availability of synchrotron radiation has been very important to the development and usefulness of the experiment. The photons in the radiation from such an electron storage ring cover the

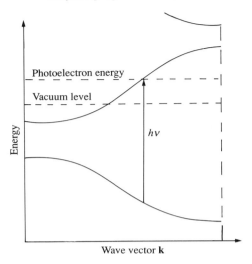

Fig. 4.12 Direct transitions from the valence band to the conduction band of solid drawn in a reduced zone scheme.

whole spectral range from the visible to the hard X-ray region, and so particular selectable wavelengths can be tuned to by using a suitable monochromator. The circular orbit of the high-energy electrons (2 GeV for instance) results in emission of radiation in a narrow cone tangential to the electron orbit with polarization in the plane of the orbit. This bright, tunable source of photons is ideal for spectroscopic studies. A discussion of synchrotron radiation and its applications can be found in Winick and Doniach (1980).

As the orientation of a single crystal sample is varied with respect to the direction and polarization of the incident radiation, so different features of the one-electron density of occupied states in the surface or the bulk appear in the experimental spectra. As in the LEED situation, electrons crossing the surface as they emerge into the vacuum conserve the component of their momentum parallel to the surface. The details of procedures to extract this density of states information from the experimental spectra are to be found, for instance, in Williams *et al.* (1980) and in Bradshaw *et al.* (1975). An example of angularly resolved photoemission spectra for the compound InSe is shown in Fig. 4.13(a), where the complexity of the detail available is very clear. However, since the energy of a peak is determined by the band structure and any selection rules on momentum in the emission process, then the spectra can be transformed into maps of the energy of the initial state versus the parallel component of the wave vector of the emitted electron. The results of such an analysis are given in Fig. 4.13(b). The dispersion of the six bands of states

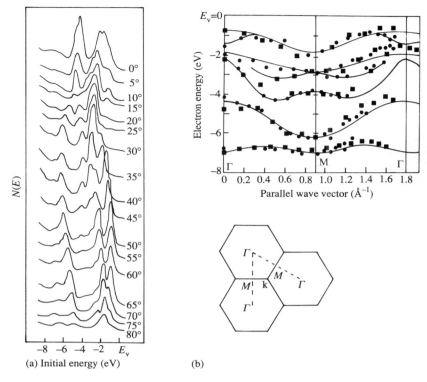

$E_v = 0$

Electron energy (eV)

$N(E)$

0°
5°
10°
15°
20°
25°
30°
35°
40°
45°
50°
55°
60°
65°
70°
75°
80°

−8 −6 −4 −2 E_v

(a) Initial energy (eV)   (b)

Parallel wave vector (Å^{-1})

0   0.2 0.4 0.6 0.8 1.0 1.2 1.4 1.6 1.8

Γ   M   Γ

Fig. 4.13 Angularly resolved photoemission spectroscopy from InSe at a photon energy of 18 eV. (a) Experimental data obtained for various polar angles of emission in the azimuth determined in the ΓMΓ azimuth in the repeated two-dimensional zone scheme for InSe. (b) The two-dimensional experimental band structure of InSe within the layers of this structure and in the ΓMΓ azimuth as interpreted from the data of (a). Squares represent data at $hv = 18$ eV and circles represent data (not given here) at $hv = 24$ eV. (By courtesy of Larsen *et al.* 1977.)

detectable in this technique is clearly revealed as is the expected symmetry of the bands about the M point.

In addition to a large body of results on the band structure of bulk solids, angle-resolved emission has been very useful in the study of surface states. Some examples are:

(1) Surface states have been identified near the Fermi level of copper and their dispersion has been measured and compared with theoretical calculations. This work is reviewed, for instance, by Heimann *et al.* (1979).

(2) The dangling bond electrons on atoms in the cleaved Si(111) surface (Fig. 4.11) have been found to give rise to a surface state (previously seen in electron loss spectra) just below the Fermi level. This peak is believed to be due

to emission from occupied states on atoms raised above the mean surface plane and disappears rapidly upon contaminating the surface. The area of the atomic and electronic structure of Si(111) is controversial. It has been described by Haneman (1987) and is discussed at greater length in a case study at the end of this chapter.

(3) Observation of the binding energies of core levels (4f) in iridium as well as core levels in gold and tungsten have revealed that the surface atoms have different binding energies from the bulk atoms—an effect which shifts the photoelectron energy by a few tenths of an eV. These shifts reflect changes in the electrostatic potential in the core region of the surface atoms which, in turn, derive from the arrangement and charge densities within the surface (e.g. van der Veen *et al.* (1980).

(4) In the case of molecular adsorbates on metal surfaces it is possible for photoemission due to some of the bonding electrons to be peaked along the molecular axis. Thus, for the much investigated system Ni(100)–CO, it has been found that one CO orbital (the 4σ) has a pronounced peak along the CO axis and that this emission peaks along the surface normal of the Ni substrate. The suggestion from this work is that the CO molecule adsorbs with its axis normal to the surface with the carbon end nearest to the metal (rather like the Pd(111)–CO example in Fig. 6.10) (Allyn *et al.* 1977).

An example of the comparison of a calculation of the electronic band structure with the results of an angularly resolved ultraviolet photoemission experiment are shown in Fig. 4.14. This is for the Al(111)–O system and used the polarization dependence that is accessible to the experimentalist when using a synchrotron radiation source. In this case the oxygen adsorbate atoms form a p (1×1) structure and have electronic states which form two-dimensional energy bands which disperse in energy as a function of the magnitude and direction of the electron wave vector parallel to the surface. This dispersion is revealed by the curvature of the energy versus direction lines in Fig. 4.14.

Plasmons

The loss processes described on p. 22 sometimes have different values for surface and bulk environments. The losses described above are single-particle excitations of electrons out of the bonds of surface atoms into empty states just above the Fermi level. They are also surface excitations, because they are specific to the special bonds which exist at the surface. Two other important kinds of excitation are into plasmons and phonons, which are, respectively, collective excitations of the electrons and the atoms in the solid. Phonons will be discussed in Chapter 5.

A plasmon is a quantized oscillation in the density of an electron gas. Such oscillations can be excited by shooting a charged particle or a photon into the

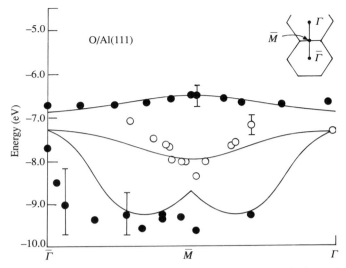

Fig. 4.14 The theoretical and experimental results for the oxygen-derived energy bands of Al(111)-O. The solid curves are the theory (by courtesy of Wang *et al.* 1981*b*) and the symbols are for the experiment (by courtesy of Eberhardt and Himpsel 1979).

solid. The Coulomb field of the former and the electromagnetic field of the latter cause a redistribution of charge in the electron gas which launches plasma oscillations. The simple theory of plasmons in a bulk solid is described in many books on solid state physics (e.g. Rosenberg 1974; Kittel 1986). The energy of a plasmon is related to the density n of a free-electron gas by

$$E = \hbar\omega_p = \hbar(ne^2/\varepsilon_0 m)^{1/2} \qquad (4.2)$$

A metal with conduction band electron densities in the range 10^{27}–10^{29} electrons m^{-3} will thus show plasmon losses of the order of a few electron volts. The exciting particle may generate more than one plasmon as it passes through the solid and so suffer multiple plasmon losses. Magnesium metal exhibits such multiple plasmon losses, as is shown in Fig. 4.15.

If a semi-infinite electron gas is terminated in a plane surface, Laplace's equation can be solved for possible charge density fluctuations and plasma oscillations predicted, which are periodic in the plane of the surface but decay away exponentially into the electron gas (e.g. Kittel 1986). The quantized unit of these oscillations is called a surface plasmon. By applying the electromagnetic boundary conditions (the tangential component of the electric field and the normal component of the displacement must both be constant across the surface–vacuum interface) and using the dispersion relationship for the dielectric constant of an electron gas (again, see Kittel 1986) it can be shown that the frequency ω_s of a surface plasmon is related to ω_p by

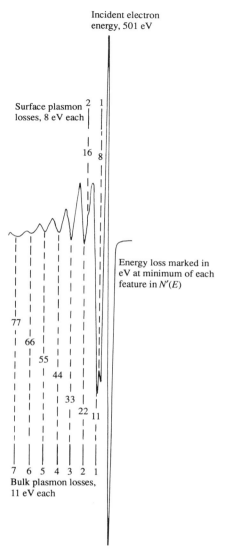

Fig. 4.15 The electron-loss spectrum from a clean surface of polycrystalline magnesium using a beam of incident electrons of energy 501 eV. The spectrum is presented in differentiated form because this reveals the multiple plasmon losses particularly clearly. Multiple bulk plasmon losses of 11 eV and multiple surface plasmon losses of 8 eV are clearly visible. (By courtesy of Dr A. P. Janssen.)

$$\omega_s^2 = \tfrac{1}{2}\omega_p^2 \qquad\qquad (4.3)$$

The surface plasmon excitations in magnesium can be seen at intervals of 7.1 eV in Fig. 4.15.

The relative intensities of bulk and surface plasmon excitations will depend, for instance, upon the energy and angle of incidence of a primary electron beam and upon the energy, angle of incidence, and state of polarization of an incident electromagnetic wave. For incident electrons the surface plasmon becomes more pronounced as the primary energy is reduced and as the angle of incidence is increased, because in both cases the penetration of the electrons decreases and so does the probability of exciting bulk plasmons. Of course, if the primary energy falls below the surface plasmon energy, no collective excitations of the electron gas are possible.

A detailed study of plasmon loss processes may help to obtain some understanding of the distribution of charge at a surface. The work function is a gross property in that it is a single parameter whole value is determined by a number of independent microscopic properties of the solid and its surface. Just as in bulk solid state physics, the densities of available and of occupied states for electrons and the relationship between electron energy and momentum (the *dispersion relation*) are useful concepts in describing surface electronic properties.

One means of obtaining an idea of surface charge distributions is to attempt to find the dispersion relationship for surface plasmons. One method for doing just this is called *inelastic low-energy electron diffraction* (ILEED). The apparatus used to perform an ILEED experiment can be the same as the LEED apparatus of Fig. 3.14, but the potentials and electronics are arranged as in Auger spectroscopy, so that electrons that have suffered a selected energy loss ΔE reach the screen and are detected. Experimentally, the variation of the collected electron current with primary beam energy, angle of incidence ϕ, azimuthal angle ψ, angle of emitted electron beam ϕ', and energy loss ΔE is observed. These rather complex data can then be compared with models for the scattering process in the solid. If processes which involve a plasmon loss before elastic diffraction and a plasmon loss after elastic diffraction are both taken into account it is possible to find the dispersion relations for both bulk and surface plasmons. This approach has been applied with some success to Al(111) surfaces by Porteus and Faith (1973).

Single atom spectroscopy and the STM

Scanning tunnelling microscopy was introduced in Chapter 3. It was mentioned there that if the tunnelling voltage between tip and surface is varied and the current–voltage characteristics are measured while the tip is maintained in a fixed position with respect to the surface, then the $I(V)$ curve so observed contains spectroscopic information about the surface. This information is an atomic scale if the tip is sufficiently sharp that there is only a single atom at its apex which is dominating the tunnelling current. This kind of spectroscopy can be carried out at each position of the tip as it is scanned across the surface, and the result is a set of spectra at each place on the surface where an $I(V)$ curve is measured. Alternatively, it can be thought of as a set of energy-resolved images of the surface, each image being obtained for a particular tunnelling voltage. As mentioned in Chapter 3 this is known as *current imaging tunnelling spectroscopy*, or CITS.

The striking differences that can be seen in STM images as the tunnelling voltage is varied are most easily observed from semiconductor surfaces because of the surface sites present in these materials and the ways in which these states are localized upon particular atoms. This was first illustrated using Si(111) by Hamers *et al.* (1986) and some of their results are shown in Fig. 4.16. These kinds of measurements, taken together with RHEED, LEED, reflection electron microscopy, and EELS measurements, and theoretical considerations as to the electronic structure of the Si(111) surface have led to the so-called *dimer adatom stacking fault* (DAS) model for Si(111)7×7, which is discussed in slightly more detail in the case study below.

Surface optics

When plane polarized light is reflected from a solid surface the reflected wave is, in general, elliptically polarized. The ellipticity can be measured using standard techniques of polarimetry (Beshara *et al.* 1969). The experimental results can be expressed in terms of the amplitudes of two components of the oscillating electric field in the reflected light and the phase difference between them. The measurement and interpretation of this ellipticity as a function of the angle of incidence, the plane of polarization of incident light and its wavelength form the subject of *ellipsometry*. It is a subject with a long history, since Malus detected polarization by reflection from metals in 1808, but it has been plagued by two substantial difficulties, which could be removed only in the relatively recent past.

The first difficulty arises because the solution of Maxwell's equations for reflection at a real (absorbing) surface leads to complicated expressions involving the refractive index n_c, the wavelength and geometrical parameters. The refractive index itself is a complex quantity involving a real refractive index n and an imaginary term due to absorption and usually called the extinction index k.

$$n_c = n - ik. \tag{4.4}$$

The values of n and k and their dependence on wavelength, are interesting physical parameters in that they are related to the response of the electrons in the solid to the electric field of the incident light wave. The real part of the

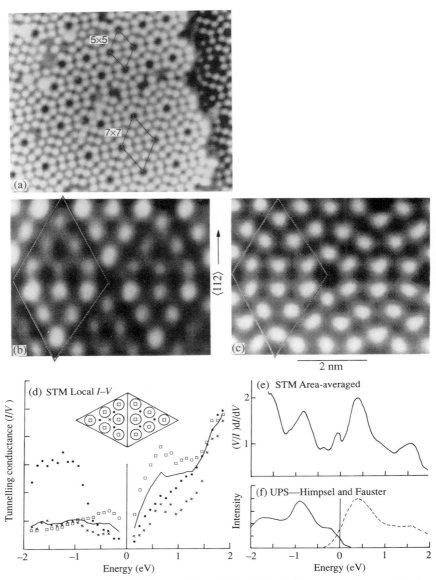

Fig. 4.16 Some STM and CITS images from Si(111). (a) An STM topographic image showing both 7 × 7 and 5 × 5 ordered regions. (b) A CITS image with the tip positive with respect to the Si(111) such that tunnelling occurs from occupied surface states in the Si to empty states in the tip. The region of the surface is a 7 × 7 area of the type visible in (a). (c) A CITS image with the opposite polarity to (b) such that electrons tunnel into empty states in the surface. The brightest spots are in different positions with respect to the corners of the unit cell than they are in (b). (d) Local I(V) spectroscopy at various sites within the 7 × 7 unit cell as specified in the inset. (e) The data of (d) normalized and averaged over the entire unit cell and compared with (f) the UPS and inverse photoemission spectra from a similar Si(111)7 × 7 surface (Himpsel and Fauster 1984). It can be seen that features in the tunnelling spectra correspond to features observed in the photoemission spectra. (By courtesy of R. J. Hammers 1992.)

refractive index is determined largely by the extent to which atoms in the solid are polarized by this electric field. The imaginary part is determined by the energy lost from the incident wave to single-particle and collective excitations of the electrons. Thus, the optical properties are intimately related to the plasmon dispersion relations mentioned above. However, the parameters actually measured in an ellipsometric experiment are related only through complicated equations to calculable quantities like the reflectivity and phase change Δ on reflection. In turn, these quantities are related through a second set of complicated equations to the values of n and k at any particular wavelength. The solution of this difficulty (which varies from simply tedious to quite hideous—see e.g. the review by Heavens (1964)) awaited the development of the electronic computer.

The second difficulty arose from the very high surface sensitivity of the ellipsometric method. UHV techniques and the methods of surface analysis described in Chapters 2 and 3 were required before reproducible results could be obtained and interpreted. Vrakking and Meyer (1971) have shown how the size of an Auger electron peak and the change $\delta\Delta$ in the phase change on reflection due to adsorbed layers of O, HS, and $CH_3 SH$ on Si(100) are linearly related up to at least monolayer coverage (depending upon wavelength). Observable changes in Δ are possible in these systems for as little as 5 per cent of a monolayer of coverage by the adsorbed species.

Electron spin resonance

Electron spin resonance spectroscopy continues to be a powerful technique for probing the bonding and environment of atoms in bulk solids. The spectrometer is designed to measure the adsorption of energy from an electromagnetic field due to changes in the spin states of unpaired electrons. Resonance occurs at a frequency v for a sample in a magnetic field H when

$$hv = g\beta H, \tag{4.5}$$

where g is the spectroscopic splitting factor and β is the Bohr magneton. If the orbital energy levels are split by a crystal field, then the size and symmetry of g will depend upon the magnitude and symmetry of the crystal field.

The technique has been applied to the study of surface atomic species (e.g. Lunsford 1972), but, in order to obtain sufficient sensitivity, it has been normal practice to study adsorption upon the surfaces of a collection of very small particles. Although increasing the surface area, this approach is almost inevitably confined to conventional vacuum techniques (as opposed to UHV), owing to trapped gases. In addition, it gives results which are some kind of sum over the effects of many different surface orientations. Nevertheless, the method is so powerful in bulk chemical investigations that means may be found to extend its application to clean surface studies.

Case study: the long search—Si(111)7 × 7

In many places so far in this book examples have been drawn from the study of
the (111) surface of clean silicon. The reconstruction of this surface to a 7 × 7
form was illustrated in Chapter 3, where both LEED and STM results were
shown, the Auger spectra of the clean surface were presented in Chapter 2, and
some electron loss spectra have been given in this chapter. The large and
concentrated effort has arisen worldwide both because of the technological
importance of this surface of silicon in the fabrication of integrated circuits and
because of the intrinsic interest in attempting to understand the beautiful
LEED and STM patterns that are observed.

   The Si(111)7 × 7 LEED pattern was first reported by Farnsworth *et al.*
(1959), and it was only during the last few years of the 1980s that the solution
to the surface structure responsible for this pattern emerged. The situation up
to about 1987 has been reviewed by Haneman (1987). Early workers were
suspicious that the Si(111)7 × 7 surface was a symptom of the contamination
of the material which was causing reconstruction by formation of chemical
bonds. One suspicion was that there was a contaminant (perhaps carbon or
chlorine) that was trapped in the material beneath the surface. Auger electron
spectroscopy could not reveal this impurity because it was deeper than the
inelastic mean free path of the Auger electrons. This is a difficult proposition to
test because it depends upon a negative result (no feature in an electron
spectrum which can be attributed to the presence of an impurity). Nevertheless
SIMS and depth profiling experiments were unable to discover any such
buried impurity, and by the mid-1970s it became widely believed that the 7 × 7
structure is a property of clean Si(111).

   The structure is created in Si(111) surfaces that have either been
mechanically cut and polished or cleaved *in situ* and then cleaned in UHV by
heat treatment at about 950°C followed by a similar shorter heat treatment at
a somewhat higher temperature and then slow cooling to 950°C followed by
rapid cooling to room temperature. The rapid cooling ensures that any carbon
which has dissolved into the silicon during the heat treatment does not
segregate to the surface at lower temperatures (for which the solubility of
carbon in silicon is lower). The 7 × 7 structure is then stable over a wide
temperature range and appears to be the stable, low-energy state of clean
Si(111).

   The use of LEED crystallography to establish the structure of this surface
was clearly a formidable task. A 7 × 7 structure in the surface layer alone must
contain at least 49 atom positions, each of which requires three coordinates to
specify a site with respect to the underlying material. If the reconstruction
involves several layers then even more coordinates have to be determined. For
example, a four-layer reconstruction may need at least 588 (49 × 3 × 4)
independent coordinates to specify the positions of atoms in the unit mesh if
there are no correlations between these positions. Of course, there are

correlations between positions due to the interatomic forces, but there may still be a very large number of independent parameters to be determined, and a LEED study requires a multiple scattering calculation for each possible model for the arrangement of the atoms in the unit cell. Tensor LEED reduces the computational difficulty because it is a fast technique for exploring perturbations in a structure that has been accurately calculated with a full dynamical model. Notwithstanding the help provided by tensor LEED, the problem remains demanding both for the measurement of sufficiently large and precise data set for this information to be determinable in principle and for the amount of computation required for the exploration of many different structural models. Other inputs are required in order to replace the number of possible structures which need to be evaluated by LEED theory–experiment comparisons. At first, such inputs were provided by calculations of simpler structures that may contain features in common with $Si(111)7 \times 7$ (e.g. $Si(111)2 \times 1$), by electron loss spectroscopy revealing surface states which could be attributed to dangling bonds, again by use of theory and by ion scattering experiments (which may conceivably have modified the surface during the experiment). This indirect approach could not resolve the problem clearly and the first breakthrough came with the invention of STM and its application to $Si(111)$ by Binnig *et al.* (1983).

A topographic STM image of a rather perfect region of the $Si(111)7 \times 7$ structure is shown in Fig. 4.16(a), where the bright spots are interpreted as being due to the silicon atoms nearest to the field emission tip. Binnig *et al.* suggested that the image contrast arose from an array of 12 adatoms in the unit mesh arranged on top of an unmodified substrate. However, close inspection of the brightness in these spots reveals an asymmetry within the unit mesh, dividing it into two halves. Theoretical work by Chadi (1984) also led to the proposal that there are 12 adatoms in the unit mesh, plus a bond breaking across the centre which might account for the asymmetry in the STM image. Later however, Yamaguchi (1985) calculated that this adatom model would lead to very high surface strain energies which meant that the model was an unlikely possibility. Other suggestions were explored, involving adatom clusters in pyramidal groups of four in the positions of the bright spots in the STM image or trimers of adatoms with a structure resembling the stacking fault apparently across the unit mesh giving the asymmetry in contrast.

The next step in the search was provided by Takayanagi *et al.* (1984) who devised methods for thinning an $Si(111)$ sample sufficiently for it to be used in a transmission high-energy electron diffraction experiment. This kind of diffraction is described with reasonable precision by kinematic scattering theory, which is substantially less computationally intensive than LEED. Further, a large number of diffracted beams from the 7×7 structure could be observed and their intensities measured. This provides an adequate database of measurements for the techniques of X-ray diffraction to be employed for a first estimate of the possible surface structure. In this case Patterson functions were used to interpret the intensities of 2760 diffracted beams. The diffraction pattern is shown in Fig. 4.17 and the model which is the outcome of this work is sketched in Fig. 4.18.

Fig. 4.17   The transmission electron diffraction pattern of the Si(111)7 × 7 structure which led to the DAS model of the crystallography. (By courtesy of Prof. T. Takayanagi.)

This model exhibits four important features:

1. Different stacking in the two halves of the unit mesh. This can be seen by comparing the sequence marked AaC on the left-hand side of Fig. 4.18(a) with the sequence AaB on the right-hand side.

2. The sides of each triangular half of the unit mesh each contain three atom pairs (or *dimers*). There are thus nine dimers around each triangular sub-mesh.

3. There are 12 adatoms over the whole unit mesh.

4. There are vacancies at each corner of the unit mesh surrounded by a characteristic 'rose' of atoms.

Many of the features seen in the STM image of Fig. 4.15(a) are seen clearly in the model of Fig. 4.18. Further, as the tip to sample potential difference is varied the empty states that are populated in the Si surface (when it is positive with respect to the tip) can be imaged, as is seen in Fig. 4.15(c), and they are

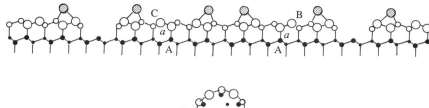

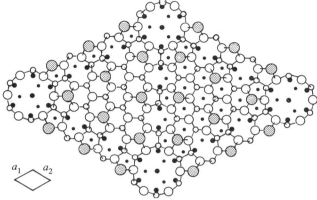

Fig. 4.18 Sketch of the dimer adatom stacking fault model (DAS) of Takayanagi *et al.* (1985). (a) View of the top few atomic layers in a plane normal to the surface and through and behind the long diagonal of the 7 × 7 unit mesh. All the atoms are Si atoms, but are drawn with different symbols to indicate their different environments. The adatoms are filled in solidly. (b) In the plan view from above the surface the symbols decrease in size as the Si atoms are deeper into the solid. (By courtesy of D. Haneman.)

quite different from the sites imaged when the tip is positive with respect to the surface and occupied surface states are being seen. The energies of these states can be measured from the $I(V)$ curves at each side and compared with theory. The adatoms of the unit mesh give rise to quite large tunnelling currents, even at low bias voltage, indicating that they have a high density of states near to the Fermi energy E_F. This is in good agreement with a rather metallic-like surface state observed in photoemission measurements. A surface state at -0.8 eV is localized on the six 'rest' atoms marked in Fig. 4.18(b). A state at -1.8 eV arises from the back-bonds between the adatoms and the first full atomic layer.

Having described such a model so completely, it could be tested using LEED as suggested in Chapter 3. This has been done by Tong *et al.* (1988). It has also been tested using grazing incidence X-ray diffraction by Robinson *et al.* (1986). Theoretical work by many authors has substantiated the low surface energy of this structure and estimated the energies of the surface states on various sites. Thus, STM, high-energy diffraction, LEED, X-ray diffraction, loss spectroscopy, photoemission studies, and theoretical calculations of crystallography and surface electronic properties have all contributed to the present acceptance of this model as a good description of the Si(111) surface.

Summary

The wave functions of the electrons of a solid are expected to be different at the surface as compared to the interior, and indeed are found to be so. This difference is important in a variety of contexts—e.g. it affects the manner in which an adatom bonds to the surface (Chapter 6), the emission of electrons into the vacuum, and the details of electron diffraction processes (Chapter 3). Both theoretical and experimental techniques for the detailed evaluation of surface densities of states and dispersion relations for surface excitations are available. Once the surface crystallography is known the electronic properties of many different kinds of surfaces can be predicted theoretically. If the crystallography is not known then the theoretical task is more substantial, because self-consistency has to be sought by allowing the atoms to move to minimize the free energy of the solid and its surface. The work function and the ellipticity of reflected light are both very sensitive to small changes in the electronic states at the surface and can be interpreted in terms of atomic models. ILEED, electron loss, and angularly resolved photoelectron spectra show features characteristic of surface states, and these methods have been developed so as to give detailed information on the electronic properties of rather large surface areas. Single atom spectroscopy is possible with STM, and this is a powerful aid to the identification and siting of localized surface states.

5
Surface properties: atomic motion

Up to this point the discussion of both methods and properties has been in terms of rigid lattices of atoms or molecules. In practice, of course, the atoms are in motion, and this motion should be included in a treatment of any properties it may affect. In this chapter the atomic vibrations at a surface will be discussed, as they reveal themselves in the temperature dependence of the intensity of a LEED pattern and in special features in electron-loss spectra. More extreme atomic motions occurring during the melting of a surface and during the motion of an adatom across it will also be outlined.

Surface lattice dynamics

It is well known that there is a fall in the intensity of the beams in an X-ray diffraction experiment (using a bulk single crystal) as the crystal temperature is raised. At the same time the intensity in the diffuse background of the diffraction pattern becomes higher. The simplest explanation of these observations is that the individual atoms of the crystal are vibrating independently about their equilibrium positions and, as a result, the exact Bragg conditions is not met. This is because scattered waves from the rigid lattice that were adding up in phase now have phase differences fluctuating with time due to the motion of the scatterers. The effect of this motion upon the intensity of the elastically diffracted beams is described in many textbooks (e.g. Kittel 1986). If I_o is the intensity elastically scattered into a beam by a rigid lattice, then the intensity I_g due to elastic scattering by a vibrating lattice in the direction determined by Bragg scattering due to a reciprocal lattice vector $\mathbf{g}$ is given by

$$I_g = I_o \exp(-\alpha \langle u_2 \rangle \mathbf{g}). \tag{5.1}$$

In deriving this equation it is assumed that the atoms are in simple harmonic motion. $\langle u^2 \rangle$ is the mean square amplitude of vibration in the direction of $\mathbf{g}$ and α is a constant whose value depends upon the number of dimensions in which the atoms are allows to vibrate. If oscillation in one dimension along $\mathbf{g}$ is chosen as a model then $\alpha = 1$; if oscillation in three dimensions is chosen, $\alpha = \frac{1}{3}$. The exponential factor of eqn (5.1) is usually called the Debye–Waller factor, and is often written as $\exp(-2M)$.

The same kind of effect is observed in LEED, but because LEED intensities arise from the first few atomic layers of a crystal the appropriate value of $\langle u^2 \rangle$ is that for the surface atoms. Indeed, as the energy of the incident electron beam is raised, the penetration increases and the relevant number for $\langle u^2 \rangle$

changes from a surface to a mainly bulk value. Because of the absence of nearest neighbours on the vacuum side it is likely that $\langle u^2 \rangle$ at the surface will be greater than that in the bulk. Further, this very asymmetry in the potential around the surface atoms is likely to require an anharmonic description of the lattice vibrations. This must be particularly true of the component of vibration normal to the surface, u_n.

By using the Debye model of the solid as a three-dimensional elastic continuum it is possible to derive theoretical expressions for $\langle u^2 \rangle$ in eqn (5.1) (Kittel 1986). This can be inserted in eqn (5.1) and a new equation derived which relates the observed intensity to other measurable quantities. A beam of wavelength λ incident upon the surface at an angle ϕ is scattered into an (00) beam whose temperature-dependent intensity $I_{00}(T)$ is given by

$$I_{00}(T) = I_{00}(0)\exp\left\{-\frac{12h^2}{mk}\left(\frac{\cos\phi}{\lambda}\right)^2\frac{T}{\Theta^2}\right\}. \tag{5.2}$$

In this equation $I_{00}(0)$ is the specularly reflected intensity from a rigid lattice, h is Planck's constant, m is the atomic mass, k is Boltzmann's constant, T is the temperature, and Θ is the Debye temperature.[†]

If the intensity of a specular LEED spot is measured at constant λ and ϕ as a function I of the temperature T then a plot of $\log\{I_{00}(T)\}$ versus T should be a straight line and eqn (5.2) can be used to derive a value for Θ. MacRae's (1964) data for Ni(110) are plotted in Fig. 5.1, which shows only his results for the specularly reflected beam. At the energy of 35 eV used to produce these results the value of Θ found from the slope of the line is 220 K. The bulk Debye temperature of nickel is 390 K, and so this result suggests that the atoms in the layers penetrated by the 35 eV electrons (the top one or two layers) have higher values of $\langle u^2 \rangle$ than do the bulk atoms. Some results of this kind for a few systems are listed in Table 5.1. The values of $\langle u^2 \rangle$ derived from such experiments depend upon the accuracy with which multiple scattering effects are included in calculating the LEED intensities and the force law used to describe the interaction between the atoms. Because Tong *et al.* (1973) chose to study Xe(111) surfaces, which show nearly kinematical LEED intensities, they could examine the effects of choosing different force laws, and it is this range of choices which gives the spread of values to Θ and $\langle u_n^2 \rangle$ in Table 5.1.

Another result of the change in the number of nearest neighbours on moving from the bulk to the surface is that different modes of vibration of the lattice become possible. The vibrational states of the bulk crystal are quantized, each quantum being called a phonon. The vibrational properties of the bulk solid are described by using a phonon dispersion relation $E(q)$ which expresses the energy E of a phonon as a function of its wave vector magnitude q. Just as there

[†]The Debye temperature is a characteristic of the bulk solid which is often first encountered in a discussion of the heat capacity. It is associated with energy of the highest frequency (ω_{max}) phonon mode possible in the Debye model of vibrations in the solid, $h\omega_{max} = k\Theta$ (Kittel 1986).

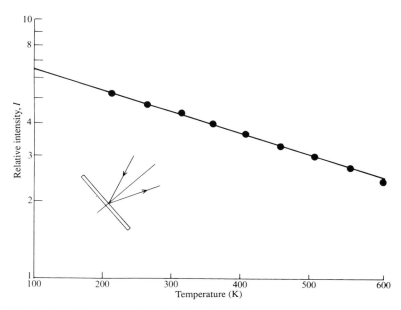

Fig. 5.1 A plot of the measures intensity I (on a logarithmic scale) versus temperature T for the specular LEED spot of Ni(110) at 35 eV. The slope yields a value of 220 K for the Debye temperature. (After MacRae 1964.)

Table 5.1 Some surface vibrational data derived from LEED observations

Material and surface	Reference	Θ(surface) (K)	Θ(bulk) (K)	$\dfrac{\langle u_n^2 \rangle \text{ surface,}}{\langle u_n^2 \rangle \text{ bulk}}$
Xe(111)	Tong *et al.* (1973)	30–35	43	3.5–2
Bi(0001)	Goodman and Somorjai (1970)	48	116	2.4
Ni(110)	MacRae (1964)	220	390	1.8
Cr(110)	Kaplan and Somorjai (1971)	333	600	1.3

$\langle u_n^2 \rangle$ is the mean square vibrational amplitude normal to the surface.

can be gaps within which particular values of k are not allowed in the electron dispersion relation $E(k)$ of a solid, so there can be gaps in $E(q)$ for the phonons. For electrons the gaps are due to $\mathbf{k}$ satisfying the condition for elastic scattering by the periodic potential of the solid. For phonons the gaps occur when q lies on a Brillouin zone boundary because atomic vibrations with wavelengths shorter than the interatomic spacing are not possible (see e.g. Rosenberg 1974).

At the surface there are vibrational modes which are not allowed within the solid. They are analogous to the surface plasmons in that they propagate with wave vector **q** parallel to the surface and decay exponentially away in amplitude along the surface normal. These modes are called *surface phonons*.

The two most widely used techniques for observing the vibrational spectra of solids are infrared spectroscopy and inelastic neutron scattering. The latter is particularly powerful in that both the energy and the momentum of a neutron which has lost energy in exciting a phonon can be measured. By varying the sample orientation and the neutron detector position, the function $E(q)$ can be determined. However, the neutron–phonon scattering cross-section is very weak, and so the technique has not been applied to surface problems. In infrared spectroscopy the electric field of the incident infrared beam can couple to the phonon modes of the solid and the resultant absorption of the beam can be measured as a function of the energy of incident beam. Spectrometers of this type are made to function over the energy range 1.5 meV–0.1 eV (wave numbers of 12–800 cm^{-1}). Sufficient sensitivity for surface modes to be detected can easily be obtained with infrared techniques if the sample is in the form of a large number of small partices. The high surface/volume ratio has thus allowed the detection of surface phonons in materials like magnesium oxide. However, such small-particle experiments are difficult to interpret, as many crystal faces are present on the particles and contamination by impurities is a strong possibility (Chapter 1). However, the energy resolution of infrared absorption experiments can be very high compared with electron loss spectroscopy (EELS)—therefore it is a very useful technique which is a powerful complement to EELS experiments if sufficient surface sensitivity can be achieved. Some applications of this kind of infrared spectroscopy to surfaces are described by Amberg (1967). The enhancement of IR absorption sensitivity to allow the study of the vibrational states of flat surfaces is outlined below.

The excitation of surface phonons by low-energy electrons, used in UHV conditions and with single-crystal surfaces has been demonstrated by Ibach (1972, 1977). He has used very high-energy resolution electron-loss spectroscopy (Chapter 4). The high resolution is essential because phonon losses are small (a few tens of meV) and are normally submerged in the broad energy distribution of electrons emitted from a conventional electron gun. This difficulty is overcome by using a dispersive analyser and an electron gun together to produce a monochromatic beam. After scattering from the solid surface the electrons are energy-analysed with a second dispersive analyser. A sketch of the experimental arrangement is shown in Fig. 5.2 and the energy-loss spectrum from a UHV cleaved Si(111) (2 × 1) surface structure obtained by Ibach is shown in Fig. 5.3.

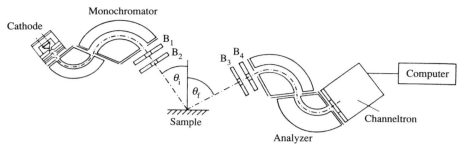

Fig. 5.2 An arrangement of electron source, monochromator, sample, and spectrometer used to obtain a sufficiently high energy resolution for phonon spectroscopy. Both dispersive elements are 127° cylindrical capacitors and the energy of the electrons incident upon the sample can be between 1 and 500 eV. (By courtesy of Rocca *et al.* 1986.)

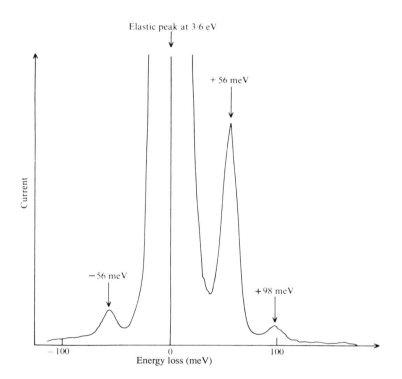

Fig. 5.3 Energy loss spectrum from Si(111)(2 × 1) using a monochromatic primary beam of energy 3.6 eV. The plane of incidence contains one of the {011} directions. The small peak at 98 meV is attributed to a contaminant. A loss at +56 meV and a smaller gain at −56 meV are visible. (After Ibach 1972.)

This method has been applied with some success to the study of molecular adsorbates upon metals. It is variously referred to by the acronyms EELS (*electron energy-loss spectroscopy*) or LEELS (*low-energy electron-loss spectroscopy*) or HREELS (*high-resolution electron energy-loss spectroscopy*). It can be used to identify the adsorbed species by recognition of the characteristic vibrational modes. If the vibrational spectrum is known from, say, infrared absorption studies of bulk material, then it is a simple matter to compare the IR spectrum of the standard and LEELS information from the adsorbate. This approach is particularly useful if it is suspected that the molecule may have dissociated when adsorbing. For example, if CO has dissociated upon adsorption then the vibrational mode corresponding to the CO molecule 'stretching' along its axis will have disappeared.

At low incident electron energies (a few eV) the electron acts as a source of electric field and, together with its image charge in a metallic substrate, produces a field normal to the surface. This field excites (dominantly) vibrations with a dipole moment normal to the surface. If the symmetry of the adsorbed molecule allows only vibrations parallel to the surface, then these will either not be excited or will be only weak.

The vibrational mode corresponding to the substrate–molecule bond will move in frequency as this bond strength varies. In this way it is possible to obtain clues about the surface site to which a molecule is bound. Thus the vibration frequency for oxygen adsorbed on a four-fold hollow site of W(100) is lower than for the same atom in a bridge site, which, in turn, is lower than on a site 'on top' of a surface W atom. This is interpreted as a sharing of the bonding force between 4, 2, and 1 nearest neighbours, respectively, with the individual bonds making decreasing angles with the surface normal. The lowest vibrational frequency is thus associated with O movements normal to the surface and an O–W bond direction nearer to the surface plane.

The dependence of the loss spectrum upon the site at which an atom is adsorbed is illustrated clearly in a diagram due to Woodruff and Delchar (1988) and reproduced in Fig. 5.4. The number of peaks in the loss (or the infrared absorption) spectrum depends upon the number of degrees of freedom of the adsorbate to oscillate in such a way that a dipole moment is created in the direction normal to the surface. The energies of the peaks depend upon the force constants in the interatomic interactions.

An example of the electron-loss spectrum of an adsorbate covered surface is shown in Fig. 5.5, which is for a Ni(100)c(2 × 2)–S overlayer. The two spectra were measured with incident electrons of energy 175 eV (Fig. 5.5(a)) and 170 eV. The plane of scattering was parallel to the side of the sulphur unit mesh for both spectra. The large peak at 112 cm^{-1} (13.9 meV) corresponds to what is known as a *Rayleigh wave*. This corresponds to a long wavelength acoustic wave propagating parallel to the surface. The smaller peaks at 316 cm^{-1} (39.2 meV) and 381 cm^{-1} (47.2 meV) are shown by Rocca *et al.* (1986) to be due to differently polarized oscillations of the adsorbate. The two spectra in Fig. 5.5

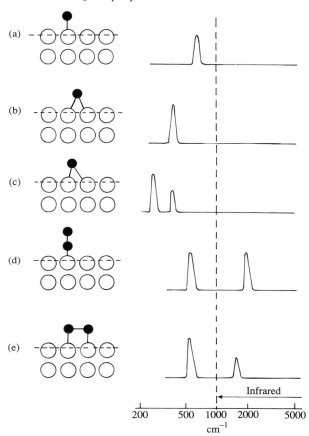

Fig. 5.4 The loss peaks to be expected from adsorbates in various sites upon a surface. (a) A single atom bonded in an 'on top' site immediately above a substrate atom. A single electron loss peak is observed corresponding to oscillations of the adatom in the direction along the surface normal. (b) A single atom in a 'bridge' site equidistant between two of the substrate surface atoms. There is still a single loss peak because vibrations creating a perpendicular dipole moment require oscillation normal to the surface. The peak is at higher energy than case (a) because there are two nearest neibours now instead of one. (c) Now a single adatom is situated asymmetrically in a bridge site. Two loss peaks are observed. One corresponds to the stretching modes of the long bond (the lowest energy mode) and the other to stretching of the shorter bond. In both cases the dominant motion of the adatom is normal to the surface. (d) A diatomic molecule adsorbed in an on top site. Two peaks are observed. One corresponds to the stretching mode of the molecule itself. The other corresponds to the oscillations of the whole molecule's distance from the surface. (e) A diatomic molecule 'prone' on the surface—i.e. the axis of the molecule is parallel to the surface. The low-energy peak arises from the vibration of the molecule normal to the surface. The high-energy peak is due to oscillation of the bond length inside the molecule.

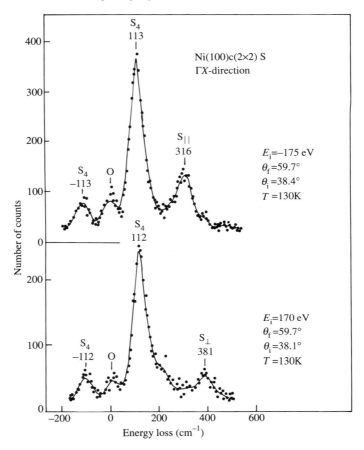

Fig. 5.5 Electron energy loss spectra for Ni(100)c(2 × 2)–S measured using a high-energy resolution spectrometer as shown in Fig. 5.2. The measurement was made at 130 K. (a) Incident beam with 175 eV. (b) Incident beam with 170 eV. The unusual units (cm^{-1}) for the electron energy loss are convenient in this kind of spectroscopy in order to aid comparison with infrared absorption experiments where the use of the wave number is traditional. (By courtesy of Rocca *et al.* 1986.)

are separated by only a small difference in the incident electron energy and yet there is a pronounced change in the detailed shape above 112 cm^{-1}. This has arisen because the scattering conditions (energies and angles) have been chosen so that only one of the modes has a significant cross-section at one of the energies.

The different modes which can be excited in the vibration of an adsorbed molecule are particularly useful in the study of simple organic adsorbates. Bond stretching and bond rotation modes can be observed and a great deal learned about the adsorption site and the orientation of the molecule in that site. More information can be found in the book by Willis *et al.* (1983).

Although sophisticated techniques are required to observe these energetically small surface phonon losses by electron spectroscopy they offer a potentially important means of obtaining information about the surface. As the surface phonon energy depends upon the type, the bonding, and the crystallography of the surface atom, comparison between such observations and the energies calculated from models of the surface should help to determine these parameters. For instance, if a LEED structure determination (Chapter 3) predicts a particular geometry of surface atoms it should be possible to calculate the surface phonon energy for simple directions within that structure and compare this with the observed energy loss. This would provide an independent test of the proposed structure.

As mentioned above, the conventional bulk technique for learning about the vibrational modes of molecules is infrared spectroscopy. The sensitivity of this technique has been enhanced to enable the study of the vibrational states of adsorbates and given the title *infrared reflection–absorption spectroscopy* (IRAS). Here an infrared beam is directed in grazing incidence at the surface under study. The wavelength of the beam is scanned and the absorption of the infrared light is measured in its interaction with the surface atoms. When the frequency of the light coincides with the frequency of a vibrational mode of the surface then light is absorbed. Thus absorption edges can be observed in the reflected intensity. This method has the advantage of the extremely high energy resolutions associated with optical techniques—typically 0.05 meV. However, the infrared sources and detectors available operate over rather limited ranges of wavelength. This contrasts with HREELS, for which a very wide energy range can be obtained but at the cost of a lower energy resolution—typically 5–10 meV.

Surface diffusion

As described above, at any finite temperature the atoms at the surface of a perfect crystal are vibrating at some frequency v_0. Thus, v_0 times a second each atom strikes the potential energy barrier separating it from its nearest neighbours. Sometimes (Fig. 5.6) the thermal energy fluctuations give the atom sufficient energy for it to leave its initial site in the surface and become an adatom in a neighbouring position of potential energy minimum. This is the simplest picture for the start of self-diffusion of an atom across a perfect surface. In practice, a real surface will contain many defects on this atomic scale (Fig. 5.7), and so there will be many different sites for surface atoms. The frequency v with which an atom will escape from a site will depend upon the height W of the potential energy barrier it has to surmount during the escape,

$$v = zv_0 \exp\left(\frac{-W}{kT}\right). \tag{5.3}$$

In this equation z is the number of equivalent neighbouring sites for the atom. This equation can be combined with a mathematical treatment of an atom executing a random walk for a time t over a mean square distance $\langle R^2 \rangle$ to give

$$\langle R^2 \rangle = Dt \tag{5.4}$$

and

$$D = D_0 \exp(-W/kT). \tag{5.5}$$

D is normally referred to as the diffusion coefficient and is usually expressed in square centimetres per second. A detailed description of the thermodynamics and the diffusion theory leading to these results can be found, starting with Blakely (1973).

Because of the variety of different sites involved on a real surface a surface diffusion experiment usually gives a measure of W that is an average over several different diffusion processes. Some processes which might be expected to occur on the surface shown in Fig. 5.7 are as follows:

(1) a single adatom (10) may hop across a terrace in jumps several lattice constants long;

(2) an adatom (8) may diffuse along the length of a ledge (5);

(3) a vacancy (9) may diffuse about by being successively filled by surface atoms.

Of course, more complex processes may also occur, and it is part of the objectives of a diffusion experiment to draw conclusions about the mechanism of diffusion by comparing observed values of D with those calculated for various mechanisms.

One way of measuring diffusion coefficients is to observe the rate of blunting of the pointed end of a field-emission tip. Such tips can be produced from either polycrystalline or single-crystal metal wires by electrochemical etching. Some results found using this method are shown in Table 5.2. Details of the method and its interpretation can be found in Ehrlich (1968).

Some particularly elegant observations of the diffusion of single adatoms upon terraces in different tips in a field-ion microscope (Chapter 3, p. 85) have been carried out by Bassett (1973). He could observe $\langle R^2 \rangle$ by vapour-depositing a single adatom on a tip and comparing micrographs before and after an experiment, such as heating the tip to a known temperature for a known time. Values of D for each temperature T could then be derived from eqn (5.4) and plots of log D versus $1/T$ used to derive values of W, the activation energy for surface diffusion. Some of the results are summarized in Table 5.3. It can be seen that there are considerable variations in W from one

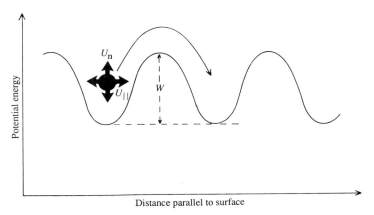

Fig. 5.6 An atom A vibrating with amplitudes u_n and $u_{||}$ normal and parallel to a simple sinusoidal potential energy surface. The probability that it will surmount the barrier of height W and reach another potential energy minimum is proportional to $\exp(-W/kT)$.

crystal face to another, even for the same type of atom on the same substrate material. Also, the diffusion can be very anisotropic on some planes—e.g. the adatom motion is along the natural channels in the W(112) and Ir(113) surfaces. Although not shown in Table 5.3, the pre-exponential factor D_0 also varies from face to face and between materials—between 3.8×10^{-7} cm^2 s^{-1} for tungsten adatoms on W(211) and 1.5×10^{-2} cm^2 s^{-1} for rhenium adatoms in W(110).

The advent of STM as described in Chapter 3 and 4 has meant that images can be obtained in which the contrast represents atomically resolved features. Therefore this technique is also a powerful means of studying directly the diffusion of single atoms and molecules over a surface.

Theoretical treatments of the diffusion of single adatoms on clean metal surfaces are difficult in that it appears to be necessary to include both the relaxation between substrate atoms and the adatom, and also the variation that can occur in the bond strength per bond when the number of nearest neighbour atoms is varied, if agreement between theory and experiment is to be obtained. Other experimental methods of studying surface diffusion do not involve the observation of single atoms but the change in shape of gross features on the surface by mass transport. One powerful technique of this kind uses a measurement as function of time and temperature of the amplitude of a sine wave topography etched chemically into a crystal surface. The amplitude of the surface roughness can be measured *in situ* in UHV by allowing light from a laser to be diffracted by the sine wave and measuring the intensity distribution in the diffracted beams. The technique is explained by Blakely (1973). The diffusion coefficients and binding energies measured with this method are different from those obtained from the FIM observations. The

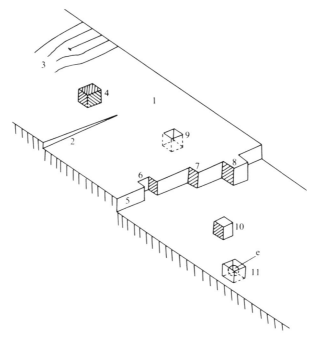

Fig. 5.7 Some simple defects that are often found in a low index crystal face. (1) The perfect flat face itself—a terrace; (2) an emerging screw dislocation; (3) the intersection of an edge dislocation with the terrace; (4) an impurity adatom (adatoms are discussed in Chapter 6); (5) a monatomic step in the surface—a ledge; (6) a vacancy in the ledge; (7) a step in the ledge—a kink; (8) an adatom of the same kind as the bulk atoms situated upon the ledge; (9) a vacancy in the terrace; (10) an adatom on the terrace; (11) a vacancy in the terrace where an electron is trapped—in an alkali halide this would be an F-centre.

latter involve the diffusion of single adatoms over a surface chosen to be free of ledges, kinks, vacancies, and impurities. The former may have at least the first three kinds of defects and, as a result, the measured activation energies of the individual processes and the populations of each type of defect taking part. Nevertheless, the mass transport techniques are of interest in so far as they help to give understanding of technologically important processes such as sintering and creep.

Surface melting

As a piece of bulk crystal is heated up, its atomic vibrations become stronger and stronger until a temperature is reached at which the crystallographic order is lost and the atoms are in a disordered conglomeration. This order–disorder transition is melting. A two-dimensional array of atoms might

Table 5.2   Activation energies for diffusion in the bulk and across surfaces of various metals measured from the blunting of a sharp tip. The activation energies for surface diffusion are often very much lower than those for bulk diffusion. It is worth recalling that room temperature corresponds to about 1/40th of an eV!

Material	Bulk diffusion activation energy (eV)	Surface diffusion activation energy (eV)
Cu	2.16	0.57
Ni	2.96	0.91
Pd	2.77	0.91
Rh(111)	1.78	—
Mo	4.01	0.3
Ta	4.29	2.61
W	6.65	2.96

Table 5.3   Activation energies for surface diffusion of single adatoms as observed by FIM. Activation energies W are in eV. Data from Bassett (1973) and Bassett and Parsley (1970)

Adatom	Terrace					
	W(011)	W(112)	W(321)	Ir(111)	Ir(113)	Rh(111)
W	0.87	0.57	0.84	—	0.99	—
Re	1.04	0.88	0.88	0.52	1.17	—
Ir	0.78	0.58	—	—	0.92	—
Pt	~0.6	—	—	<0.41	0.69	—
Rh	—	—	—	—	—	0.24

be expected to melt at a lower temperature than a bulk lattice because of the smaller number of nearest neighbours in the former. Although it is not possible to study a two-dimensional layer of atoms in isolation, it is possible to grow two-dimensional layers of atoms of one metal upon a clean surface of a second metal. Some examples of such systems showing monolayer growth were mentioned in Chapter 2 because of their usefulness in calibrating electron spectroscopic techniques such as AES, XPS, and UPS. They can also be used to study melting processes by observing their LEED patterns and the drop in intensity of LEED spots with increasing temperature.

One system which appears to show a surface melting point below that of the

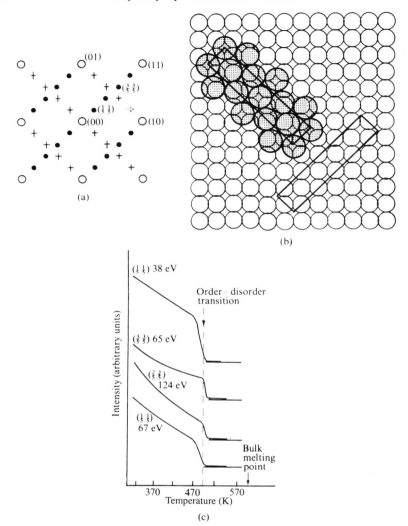

Fig. 5.8   (a) The Cu(100)c(5 × 1)–R45°–pB LEED pattern. ○: Cu substrate spots; ●: spots due to one c(5 × 1) domain of lead; × : spots due to the other domain of lead. (b) A possible model of a lead structure that would give the diffraction pattern of (a). The two domains with equivalent orientations with respect to the substrate (enantiomorphs) are indicated. (c) Temperature dependence of the LEED intensities of spots due to the lead overlayer. The transition is the sharpest observed in the copper–lead system. The curves are reversible with temperature. The zeros have been displaced for clarity of display. (After Henrion and Rhead 1972.)

bulk is monolayers of lead on various copper surfaces. This system was reported by Henrion and Rhead (1972). One example from their results is an ordered structure corresponding to a monolayer of lead in dense (111)-like packing upon the (100) surface of copper. From the LEED observations this pattern is called Cu(100)c(5 × 1)R45°–Pb (Fig. 5.8(a) and (b)). All the spots in the LEED pattern can be accounted for by recognizing that there are two equivalent orientations of this lead structure upon the copper, as shown in Fig. 5.8(b). Upon heating this structure there is no change in the lead Auger signal at 93 eV up to 673 K, from which it can be concluded that there is little or no inter-diffusion or alloying of the lead with the copper. However, the intensities of LEED spots due to the lead overlayer pass through a sharp drop which has a point of inflection at 498 K. The melting point of lead is 600 K. This drop is shown in Fig. 5.8(c). Surface melting of clean bulk lead has also been studied using medium-energy ion scattering (MEIS) by Pluis *et al.* (1990).

Summary

Atomic vibrations in a solid surface can be measured using the temperature dependence of the intensity of LEED spots in a way analogous to the Debye–Waller correction applied in X-ray diffraction experiments. The modes of vibration can be detected using a very high-energy resolution electron spectroscopy or infrared absorption spectroscopy. It is found that the atomic vibrations have larger amplitudes at the surface than in the bulk, particularly in the direction of the surface normal. Also, surface vibrational modes known as surface phonons can be found, and have energies of the order of a few tens of meV. The detection and identification of these modes is extremely useful to help identify the site at which an adsorbed atom or molecule is situated and the orientation of an adsorbed molecule with respect to the surface normal. More extreme atomic motions occur when the atoms leave their equilibrium lattice sites and diffuse over the surface or assume a disordered array. Surface diffusion of single adatoms can be observed with FIM and STM, and activation energies of the order of 1 eV are found to apply. Surface melting and surface phase changes can also occur and can be detected by LEED, angle-resolved XPS, or MEIS.

6
Surface properties: adsorption of atoms and molecules

A wide variety of events can occur when a molecule impinges upon a surface. It may be specularly reflected with no loss of energy, or it may suffer a redistribution of momentum and be diffracted by the surface, again with no loss of energy. Alternatively, and more usually, it will lose energy to the atoms in the surface by exciting them vibrationally or electronically. If it loses only a small amount of its energy and does not become bound to the surface it may be inelastically reflected. On the other hand, it may lose sufficient energy to become effectively bound to the surface with a strength that will depend upon the particular kinds of atoms involved. If this occurs the molecule is said to be *accommodated* by the surface—it has an energy appropriate to the temperature of the surface and has become *adsorbed*. As discussed in Chapter 5, it may diffuse about the surface until it picks up enough energy from thermal fluctuations to leave again or desorb. An ensemble of adsorbed molecules is called an adlayer and the average time of stay of a molecule upon the surface is called the mean stay time. Of course, even more complex events can also occur. For instance, an impinging molecule may dissociate before it can be adsorbed upon the substrate—a process known as dissociative adsorption. Adsorption of this kind is discussed at greater length by Bond (1974).

The experimental methods used to study these events can be microscopic or macroscopic. For example, a microscopic study might involve observations of the angular and energy distributions of atoms which had been scattered out of a monatomic and monochromatic incident beam. Such an experiment would be interpreted in terms of the interaction of individual atoms with the surface. A macroscopic example is the measurement of the density of the adlayer (the *coverage*) as a function of the pressure of the absorbing species and of the surface temperature. This experiment might be interpreted in terms of an average binding energy for the whole surface, which may or may not be useful for yielding information on an atomic scale.

Adsorption is of practical significance in a wide variety of problems and processes. It is the first stage in the formation of an oriented overgrowth—epitaxial growth of thin films (p. 174). It is important in catalysis, where different adsorbates upon a surface may help or hinder the progress of a chemical reaction, and it is important in vacuum technology, where it is used for pumping gases out of chambers (e.g. cryogenic pumps) and is a nuisance if adsorbed gases have to be removed in order to improve the ambient pressure. Reviews of the use of the techniques of surface science in the study of adsorption in metallurgy, microelectronics, heterogeneous catalysis, corrosion science, and polymer technology can be found by various authors in the book by Briggs and Seah (1990).

The wide range of motivations for studying adsorption, the availability of microscopic and macroscopic methods, and the large variety of surface events possible all conspire to make this area of surface studies into a broad and developing subject; only a few aspects can be touched upon here.

Some thermodynamics

In general, adsorption includes the binding to the surface of an atom or molecule arriving either from the vapour outside the solid or from the interior by processes of diffusion. The macroscopic description of the processes involved is provided by the subject of surface thermodynamics, which was developed early in this century (Gibbs 1928). The subject is developed with particular emphasis upon surfaces in the book by Blakely (1973).

In general, there are three macroscopic properties of a surface of a solid containing more than one chemical element. These are the *surface tension*, the *surface stress*, and the *specific surface free energy*. The surface tension γ is defined as the reversible work done in creating a unit area of new surface at constant temperature, volume, and total number of molecules. Under conditions of constant temperature and volume the work done in creating a new unit area of surface is the Helmholtz free energy for that area and this is defined as being the specific surface free energy. The surface stress, on the other hand, is a more complex quantity because it involves the work done in deforming a surface. A solid (usually made up of a periodic arrangement of atoms in a crystal lattice) will not deform isotropically in response to an isotropic strain, and therefore the surface stress is a tensor. Blakely describes how the surface stress tensor g_{ij} is related to the surface strain ε_{ij} and the surface tension by the equation

$$g_{ij} = \gamma \delta_{ij} + \left(\frac{\partial \gamma}{\partial \varepsilon_{ij}} \right), \tag{6.1}$$

where δ_{ij} is the Kronecker δ.

When the atoms in a surface are extremely mobile (e.g. a solid at high temperatures or a liquid) then the surface tension is independent of any deformation and the second term on the right-hand side of eqn (6.1) is zero. In this case the stress is isotropic and the values of g and γ are scalar and identical. This is not true for most situations at the surface of a solid because γ does vary with the strain—some directions in a crystal are more easily deformed than others.

Experiments that reveal the variations of some of these thermodynamic quantities are quite common in surface science. Examples will be discussed below in which the amount of an adsorbate (the coverage) is measured at constant temperature as a function of the partial pressure of the same kinds of atoms in the vapour above the surface. The plots of coverage versus pressure resulting from such an experiment are known as *adsorption isotherms*. Since the adsorbed atoms change the specific surface free energy this experiment can be interpreted using the machinery of thermodynamics. The shape of the isotherms depends upon the energy of the bonds between the adatoms and the substrate and so a microscope quantity can be estimated from the result of a macroscopic measurement.

Adsorption processes

In the case of physical adsorption (or *physisorption*), an adsorbed molecule is bound to the surface via a rather weak van der Waals type bond. This bond involves no charge transfer from the substrate to the adatom or vice versa. Rather, the attractive force is provided by the instantaneous dipole moments of the adatom and its nearest neighbour surface atoms (see e.g. Kittel 1986). The interaction can be described by the potential energy diagram shown in Fig. 6.1. An incoming molecule with kinetic energy E_k has to lose at least this

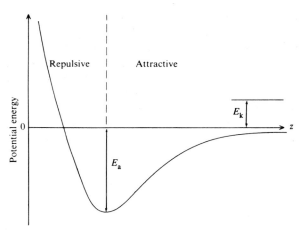

Fig. 6.1   Potential energy of an adatom in a physisorbed state on a planar surface as a function of its distance z from the surface.

amount of energy in order to stay on the surface. It loses energy (accommodates) by exciting lattice phonons in the substrate, and the molecule then comes to equilibrium in a state of oscillation in the potential well of depth equal to the binding or adsorption energy E_A. The kinetic energy associated with these oscillations is that appropriate to the substrate temperature. In order to leave the surface the molecule must acquire enough energy to surmount the potential energy barrier E_A. The desorption energy is thus equal to the adsorption energy.

The binding energies for physisorbed molecules are typically 0.25 eV or less. Such bonds are found in the adsorption of inert gases upon metals and glasses. If τ_0 is the period of a single surface atom vibration in the well of depth E_A (Fig. 6.1) then the stay time τ of this atom upon the surface is given by

$$\tau = \tau_0 \exp(E_A/kT). \tag{6.2}$$

The times τ_0 are usually of the order of 10^{-12} s, and eqn (6.2) can be used to show that stay times greater than about 1 s will not occur until $E_A \geqslant 28\ kT$. Thus, an adsorption energy of 0.25 eV will give $\tau > 1$ s only below temperatures of about 100 K. Equation (6.1) is simply related to the frequency with which an atom can escape from a potential well described in eqn (5.3). The derivation of eqns (5.3) and (6.1) is given by Frenkel (1946) and, in a more recent reference, by Weston (1985).

It is more usual for electron exchange to occur between an adsorbed molecule and the surface, in which case a rather strong bond is created with the surface and the molecule is said to be *chemisorbed*. The most extreme case of chemisorption occurs when integral numbers of electrons leave the adsorbed molecule and stay on the nearest substrate atom (or vice versa). This would be a pure ionic bond. More usually there is an admixture of the wave functions of the valence electrons of the molecule with the valence electrons of the substrate into a new wave function. The electrons responsible for the bonding can then be thought of as moving in orbitals between substrate and adatoms, and a covalent bond has been formed. A simple example of the potential energy diagram for chemisorption is shown in Fig. 6.2. Some of the impinging molecules are accommodated by the surface and become weakly bound in a physisorbed state (also called a precursor state) with binding energy E_p. During their stay time in this state, electronic or vibrational processes can occur which allow them to surmount the small energy barrier E_c (activation energy $E_c + E_p$) and electron exchange occurs between absorbate and substrate. Each adatom now finds itself in a much deeper well E_A. It is chemisorbed. The range of binding energies (heats of adsorption) in chemisorption is quite large, extending from about 0.43 eV for nitrogen on nickel to about 8.4 eV for oxygen on tungsten.

The theoretical description of chemisorption is very complex and, as yet, far from complete (e.g. Norskov 1990). However, a clue as to one way in which an adatom may change from a physisorbed to a chemisorbed state can be obtained by thinking about the processes that can occur as an atom approaches a simple free electron-like metal surface (Fig. 6.3). If the ionization energy I of the adatom is less than the work function ϕ of the metal then an electron will be transferred to the metal and the adatom becomes ionized. An example of such a process is caesium ($I = 3.87$ eV) on tungsten ($\phi \sim 4.5$ eV). On

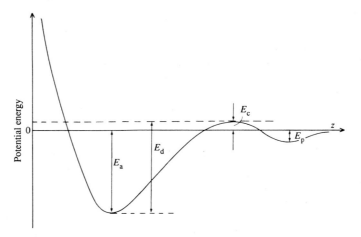

Fig. 6.2   A simple version of a potential energy diagram for chemisorption on a planar surface. Note that, once chemisorption has occurred, the desorption energy E_d is greater than the adsorption energy E_A. The potential wells contain discrete energy levels which correspond to the allowed vibrational states of the adatom. Strictly, two diagrams should be used, because the system changes once chemisorption has occurred.

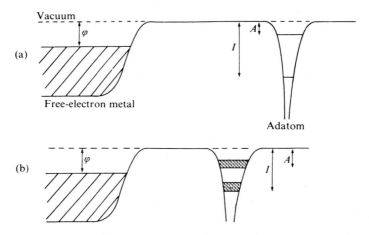

Fig. 6.3   (a) Electronic levels for an adatom so far from the surface of a free-electron metal that the levels of neither are disturbed. (b) The adatom is now closer to the metal and interacts weakly with it. The affinity and ionization levels become broadened into bands. The ionization potential I is defined as the energy required to remove an electron from an orbit and place it at rest, outside the atom. The electron affinity A is defined as the energy gained in taking an electron at rest just outside an atom and placing it in a vacant orbit.

the other hand, if the electron affinity A of the adatom is greater than ϕ an electron will be transferred from the substrate to the adatom. An example of this process might be a fluorine atom ($I = 3.6$ eV) on caesium (~ 1.8 eV)—the compound caesium fluoride is formed. The third case occurs of $I > \phi > A$, for here the atom is stable in its neutral state. Hydrogen, for instance, has $I = 13.6$ eV and $A = 0.7$ eV, and so may be expected to form neutral bonds with most metals ($\phi = 4$–6 eV).

This kind of argument depends upon the notion that the electron levels of the adatom and the metal are not disturbed by each other. Of course, this is not strictly possible. When the adatom approaches the surface it first interacts weakly with the metal and electrons can tunnel between adatom and solid. This process will shift and broaden the adatom states (Fig. 6.3(b)) and both the affinity and ionization levels become narrow bands. The tails of both these bands may overlap the Fermi level of the metal and so both affinity and ionization states may be partially filled. As the adatom approaches more closely, entirely new electronic structure may appear as the strong interaction between solid and adatom modifies the electronic levels of both.

As more atoms or molecules arrive at the surface an adlayer begins to form. If the rate of arrival of particles is J then the surface concentration at equilibrium, measured in atoms per square centimetre, is related to the stay time by

$$\sigma = \alpha_c J \tau, \tag{6.3}$$

where α is the *condensation coefficient*. It is simply the probability that an impinging particle will be accommodated upon the surface. More easily measured is the sticking coefficient S, which is the rate of increase of coverage θ with total exposure M to impinging particles,

$$S = \delta\sigma/\delta M, \tag{6.4}$$

where

$$m = \int j\,\mathrm{d}t \tag{6.5}$$

Both S and σ can vary with M in ways which depend upon the details of the interaction between adatoms and substrate and upon the topology. Some examples of what can occur in practice are indicated in Fig. 6.4, where the coverage is expressed as a fraction of a monolayer instead of in atoms per square centimetre.

Because it increases coverage, the arrival of more adatoms reduces the distance between nearest neighbours adatoms, and interactions between them become important. One effect of this interaction can be to order the adatoms crystallographically. Such ordered adsorption is usually an example of the beginning of epitaxial growth.

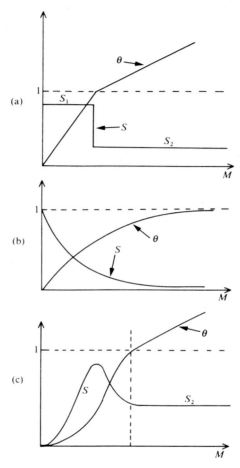

Fig. 6.4 Some plausible variations of sticking coefficient S and coverage θ with exposure M. (a) The adlayer forms on the substrate with constant sticking coefficient S until a monolayer ($\theta = 1$) is formed. Then a new value of S—that of adsorbate growing upon adsorbate—is applicable. Such behaviour is seen for some metals condensing upon others (e.g. silver on nickel). (b) Incident atoms impinge directly upon unoccupied adsorption sites and fill them. As coverage increases the number of available sites falls and so the sticking coefficient falls. Finally, a monolayer is formed and no further impinging atoms stick. Such behaviour might be shown by gas adsorption on a metal. (c) Here the adsorption site is the perimeter of a nucleus of several adatoms in a cluster. As the clusters grow their perimeters increase in length and S increases. Then they begin to touch and coalesce and S decreases again. Finally the value of S that is pertaining to adatoms growing upon the adsorbate. Such behaviour can be exhibited by metals condensing upon alkali halides.

Theory of chemisorption

The theoretical description of the processes of chemisorption inevitably draws upon many different parts of physics and chemistry. The formation of a bond between an adsorbed atom or molecule and the surface of a solid involves the participation of the least tightly bound electrons in the system. Thus, a necessary and important part of the theory must be a description of the electronic structure of atoms and molecules when they are situated just outside the surface. This is almost always more complex than both the atomic theory required to describe the properties of an isolated entity, such as a lone atom, and the solid state theory required to describe the electronic properties of a surface. Nevertheless, it is a quest worth the effort, not least because the ability to describe the physical concepts that underlie the huge weight of experimental observations helps to understand the patterns of behaviour that are observed and to assist in the prediction of new interesting or useful phenomena. The theory that can be used to do this for simple situations has been reviewed by Norskov (1990).

As outlined in Chapter 4, the theory of the electronic states at a surface often proceeds either by a large *ab initio* calculation which contains approximations about the description of the electron–electron interactions or by the construction of a relatively simple model which attempts to include only the essential parts of the physics of the system. The virtue of the latter approach is that it can at least help to provide some understanding of the limiting situations that may occur. The most important theory of this kind is known as the *Newns–Anderson* model because it was developed by Newns (1969) from a proposal due to Anderson (1961). There are two limiting cases which emerge from the application of this model to a system consisting of an adsorbate with a single valence state on a metal surface. These two extremes are sketched in Fig. 6.5. In Fig. 6.5(a) is the case where the density of states in the metal is very

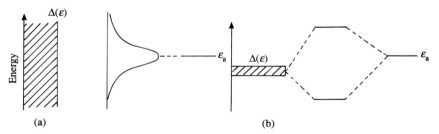

Fig. 6.5   The local density of states at an adsorbate in the Newns–Anderson model. (a) The single level in the isolated adsorbate atom is broadened into a Lorentzian distribution when the chemisorption is weak. (b) The same level is split into two states when the interaction with the metal is strong compared to the width of the metal conduction band. (By courtesy of J. K. Norskov.)

flat (i.e. the density of occupied states in the conduction band is almost independent of energy) and the adsorbate state is broadened from a single level in the isolated atom to a Lorentzian distribution centred about the original energy level. This is *weak chemisorption*. The other extreme occurs when the electrons in the metal are confined to a band which is narrow compared to the coupling between the atom and the metal. This is indicated in Fig. 6.5(b) where it can be seen that the single level in the isolated atom has split into two states—known as *bonding* and *anti-bonding*.

These kinds of behaviour have been observed in the results both of large *ab initio* calculations and of experiments. An example of weak chemisorption is provided by the photoemission spectrum of oxygen on aluminium (a nearly free electron-like metal with a flat conduction band density of states). An example of strong chemisorption is the case of CO molecules in Ni(111)—in this case the conduction band of the transition metal substrate is not all free electron-like and the splitting of the CO valence level of the molecule can be observed using angularly resolved photoemission spectroscopy in synchrotron radiation. The details of these cases are discussed by Norskov (1990).

Experimental observations of chemisorption

In order to obtain a more detailed understanding of the interaction of an adatom and a substrate it would be useful to be able to describe mathematically the shapes of the curves in Figs. 6.1, 6.2, and 6.3 and to be able to place the electronic states in a one-electronic diagram for the system. The most commonly used observables for this characterization of chemisorption are the change in work function $\Delta\phi$ on adsorption and the heat of adsorption (or desorption) to glean information about the depth of the well in Fig. 6.2; the angularly resolved electron spectra to test the electronic state of the adsorbate; a LEED or grazing X-ray diffraction experiment to determine the surface structure and therefore the position of the well in Fig.6.2; and the vibrational spectrum using phonon spectroscopy to determine the curvature of the bottom of the well. To obtain information on this adatom–solid force law is difficult, but feasible, as outlined in the following paragraphs.

Work function changes

As described in Chapter 4, is an adatom donates an electron to the conduction band of the substrate the work function decreases on adsorption. Conversely, electron transfer to the adatom increases the work function. Thus the sign of $\Delta\phi$ immediately gives information about the direction of charge transfer. In many circumstances the adatoms are merely polarized by the surface attraction. In this case they may be thought of as being polarized normal to the surface. If the positive pole is at the interface (the negative pole on the vacuum side) the work function of the substrate will be increased and the opposite will occur if the negative pole is at the interface.

If the substrate is an insulator or a semiconductor, then the density of surface states may be sufficiently large that it is these that will control the adsorption, and not the underlying band structure.

Flash desorption

The depth E_d of the well in Fig. 6.2 can be measured by using the technique of flash desorption. The sample is heated up rapidly in a chamber of known volume V and the partial pressure of the absorbate measured as a function of time with a sensitive mass spectrometer (e.g. Gomer 1967). As the temperature of the substrate passes through the energy corresponding to E_d, a sharp burst of pressure due to desorption of the adatoms is observed. From measurements of the desorption rate as a function of temperature a value for E_d can be derived.

Atomic and molecular beam experiments

Another technique for obtaining values of E_d is to measure the mean stay time τ of an adatom as a function of substrate temperature T and then use eqn (6.1). Hudson has developed some elegant methods for making such a measurement (e.g. Sandejas and Hudson 1967), and one of his experimental arrangements is sketched in Fig. 6.6. By means of a shutter in the well-collimated, monochromatic beam of incident atoms (cadmium in the case shown) the substrate can be exposed to a sudden flux of impinging atoms. Atoms leaving the surface after a mean time are detected by a mass spectrometer. The spectrometer signal can be observed on an oscilloscope as the incident beam is periodically 'chopped' by the shutter. The value of τ is derived from the exponential rise and decay of the signal due to the desorbing atoms. An example of a plot of τ (on a logarithmic scale) versus $1/T$ for cadmium on clean polycrystalline tungsten is shown in Fig. 6.7. The slope of these plots gives the values of E_d indicated beside the lines. It appears that, in this case, the first monolayer or so of cadmium is bound tightly to the tungsten surface and subsequent cadmium atoms are bound more loosely. Detailed interpretation of the magnitude of E_d for each phase is often difficult, however.

The whole field of the elastic and inelastic scattering of atoms and molecules at surfaces is being alluded to here for the first time and, somewhat as an aside, should be mentioned. If light neutral particles with energies in the thermal range (10–100 meV) are directed in a beam towards a solid surface then they do not have enough energy to penetrate into even the first layer of ion cores. Rather, any scattering is due to the interaction of the atoms of the beam with

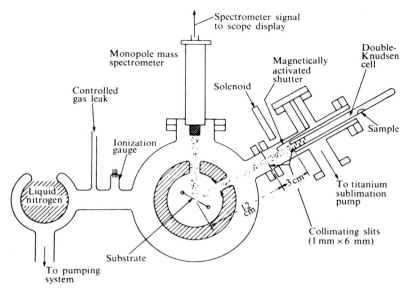

Fig. 6.6   A molecular beam experiment to study the adsorption of cadmium upon polycrystalline tungsten (Sandejas and Hudson 1967). The temperature of the sample of cadmium controls the flux of cadmium atoms falling upon a set of collimating slits. The temperature of the collimating slits controls the mean velocity of cadmium atoms reaching the substrate. The liquid nitrogen dewar around the substrate ensures that atoms scattered from the substrate do not subsequently reach the detector unless they pass directly through the detector aperture. (From Sandejas and Hudson 1967.)

the outermost part of the potential of the most exposed atoms of the solid. This part of the interaction potential is dominated by the van der Waals contribution which is rather long-range in its effects. The periodicity of a crystalline surface results in a rather weak *corrugation* in this interaction potential. Nevertheless, elastic scattering (diffraction) of He atoms from alkali halides and from silicon surfaces has been observed. Inelastic scattering is also a possibility because the ingoing atoms can excite phonons in the surface. Phonon spectroscopy has been carried out, for example, using He, Ne, Xi, H_2, or D_2 from a variety of alkali halide or metallic surfaces. Diffraction of atomic beams is another useful probe of the surface crystallography, complementing the methods described at greater length in Chapter 3. Phonon spectroscopy with atomic beams is complementary to the high-resolution electron-loss spectroscopy described in Chapter 4. Since the dispersion of the phonon modes of a surface is dependent upon the symmetry and strength of the binding between the atoms, this spectroscopy is an indirect probe of the chemisorption processes. The subject has been described by Rieder, Cardillo, Toennies, and others in the book edited by Benedek and Valdusa (1982).

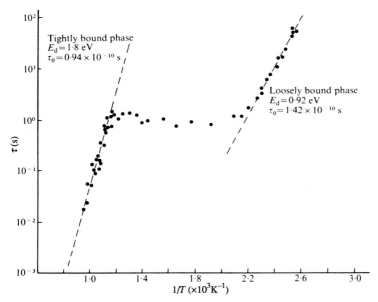

Fig. 6.7 Mean stay times for the adsorption of cadmium on polycrystalline tungsten. (From Sandejas and Hudson 1967.)

Field-electron emission

Insight as to the energy levels of adatoms (Fig. 6.3(b)) has been obtained for hydrogen and deuterium on W(100) and W(110) by energy analysing field-emitted electrons from these tungsten surfaces as a function of the coverage of adsorbate. The experimental arrangement is to combine a field-electron emitting tungsten tip (Chapter 4) and a high energy resolution electron energy analyser (the CHA of Fig. 2.3(d)). This sophisticated combination has been used by Plummer and Bell (1972). If there are empty states in a narrow band of the adsorbate–substrate combination (e.g. A to I in Fig. 6.3(b)) then electrons can tunnel from the substrate through the adsorbate without energy loss. This process is called elastic tunnelling resonance. On the other hand, field-emitted electrons may excite electronic or vibrational states of the adsorbate–substrate complex and, in so doing, lose energy. Such a process is called inelastic tunnelling. By comparing the energy distributions of field-emitted electrons from the clean tungsten surface with those at various coverages of adsorbate these two processes can be identified and something learned about the electronic states of the adsorbate. An example due to a single barium adatom upon a W(111) plane is shown in Fig. 6.8. A combination of the results obtained in this way with flash adsorption and work function measurements should provide a very powerful means of understanding the character of the adatom–substrate bond.

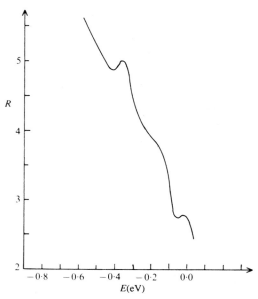

Fig. 6.8   The energy distribution of field-emitted electrons passing through a single barium atom upon a W(111) face. The vertical scale is the enhancement factor R which is the ratio between the energy distribution after adsorption and the energy distribution of electrons emitted from the clean W(111) surface. (After Plummer and Young 1970.)

Electron spectroscopies

Since the emitted electrons in both AES and XPS have kinetic energies which depend upon the binding energies of their parent states in the atoms whence they came, these spectroscopies provide key tools for the study of chemisorption. The importance of these methods is spelled out in more detail in Chapter 2. The *chemical shifts* in the energy of an XPS or AES peaks from a particular kind of atom as it is moved from the gaseous state to the adsorbed state are particularly useful in the study of adatom binding. In addition, the angularly resolved photoelectron emission spectra described in Chapter 4 form a second, and key, set of methods for the determination of the electronic structure at a surface. Understanding of this electronic structure is essential if the chemical state of surface atoms is to be understood. The reader is referred to the book by Zangwill (1988) for an introduction to this area.

In addition to the information yielded by all the methods described above it is important to realize that not only do the adatoms interact with the substrate atoms but they also interact amongst themselves. If the substrate is a single-crystal surface then it is very common for atomically ordered adlayers to form. The unit mesh (Chapter 3) of such an adlayer is determined by the interplay of the bonding with the substrate and the bonding between the atoms. These ordered adlayers have been observed using LEED in hundreds of different adlayer–substrate combinations and some are tabulated by Somorjai (1972) and by Watson (1987). There is, at present time, no general theory which enables a prediction to be made of the size and shape of the unit mesh of adatoms that will be observed for a chosen adlayer–substrate combination. However, among the observations some patterns can be seen which give clues as to tendencies. Somorjai (1972) points out that adsorbed atoms or molecules of monolayer thickness tend to obey three 'rules':

1. They tend to form surface structures in which the adatoms are close packed. Thus, they grow with the smallest unit mesh permitted by the dimensions of each adatom, the adatom–adatom, and the adatom–substrate interactions. An example of this effect is shown in Fig. 6.9, which is for Ni(100)c2–O—a more closely packed arrangement is not possible because of the large size of the adsorbed oxygen ions.

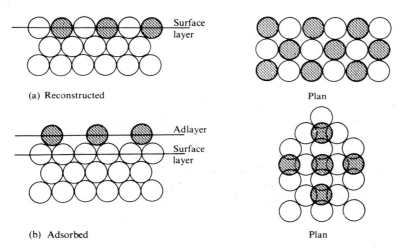

Fig. 6.9 Two possible surface structures to explain Ni(100)c(2 × 2)–O. Oxygen atoms (or ions): open circles, nickel atoms (or ions): hatched circles. (a) Reconstructed surface. Each oxygen and each nickel ion in the surface need not have the same charge provided that the whole layer is charge neutral. (b) Classical adlayer with no reconstruction.

2. They tend to form ordered structures with the same rotational symmetry as the substrate.

3. They tend to form ordered structures with unit mesh sizes rather simply related to the substrate mesh size. Thus (1×1), (2×2), c(2×2), or

(3×3)–R30 are all commonly observed. It is useful to draw out this last mesh using the information given in Chapter 3 because it is a particularly simple arrangement of the adatoms with respect to the substrate atoms.

A major difficulty associated with trying to understand and apply these 'rules' arises when an attempt is made to decide whether or not the adatoms are situated in a separate layer on top of the substrate surface. If they are, then the 'rules' can be understood in terms of rigid-ball models of adlayer and substrate surface. If they are not, then presumably the adatoms are incorporated into the surface layer, place-exchange of substrate atoms and adatoms has occurred, and the surface is said to be reconstructed. Reconstruction must be the beginning of the formation of a compound of the adatoms and the substrate atoms and for this to occur charge exchange is necessary between the atoms. Under such circumstances billiard-ball models may be dangerously misleading. Two examples of chemisorption will be discussed below with a view to illustrating both the points made above and the techniques described in earlier chapters.

Ni(100)–O   The bulk oxide of nickel—nickel oxide (NiO)—is a refractory material that melts at 2200 K. This suggests that the oxygen–nickel bond is very strong—a suspicion borne out by the value of about 2 eV for the desorption of oxygen from an ordered layer on Ni(100) (Brennan and Graham 1966). Also, NiO has the NaCl crystal structure and is a very ionic material, from which a simple view might be that it is made up of Ni^+ and O^- ions. Thus, it might be anticipated that as oxygen molecules arrive at a Ni(100) surface they would be dissociated to oxygen atoms, ionized to form O^- ions, and so bound by ionic attraction. As the coverage of O^- ions increases the Coulomb repulsion between O^- ions increases until a point will be reached where a charge neutral layer consisting of an equal number of Ni^+ and O^- ions will have a lower energy than a layer of only O^- ions. Unfortunately, such simple-minded arguments are immediately confounded by the fact that it costs energy (6.8 eV) to produce O^- from O as well as costing energy to lift the two electrons from the top of the conduction band of Ni to form Ni^+ (the work function of Ni(100) is about 4 eV). This difficulty is resolved by including the fact that a lattice of ions creates an electrostatic crystal field which must be included in evaluating the energy of the system. An O^- ion may not be stable when isolated but can easily be stable if surrounded by attractive Ni^+ ions. A careful set of considerations of this type is given by Carroll and May (1972) who conclude that reconstructed adlayers containing nickel and oxygen ions will be energetically favourable but that the oxygen ions need not necessarily carry exactly two electronic charges each.

LEED observations indicate that two stable phases exist for oxygen on Ni(100). At a coverage of a quarter monolayer there is a Ni(100)–p(2×2)–O LEED pattern and at a coverage of a half monolayer there is a

Ni(100)c(2 × 2)–O LEED pattern. The latter is of particular interest in that similar diffraction patterns are observed for the other adsorbates and substrates (e.g. Ni(100)c(2 × 2)–S, Cu(100)c(2 × 2)–O). The question of great interest is: are the oxygen atoms in this half-monolayer phase adsorbed on the surface or are they incorporated into the substrate by reconstruction of the surface layer? Two possible models for these alternatives are shown in Fig. 6.9. One way of choosing between these models is to carry out a calculation of LEED $I(V)$ curves (as described in Chapter 3) for each model and compare these theoretical curves with experimental $I(V)$ determinations. This has been done (see Pendry 1974), and the conclusion at present is that such comparisons suggest that a classical adlayer is formed (Fig. 6.9(b)). At the same time, considerations of heats of adsorption and Auger electron spectra (see the discussion in Carroll and May (1972)) suggest a reconstructed surface like Fig. 6.9(a) with mixed ionic and covalent bonding. This is supported by RHEED observations due to Garmon and Lawless (1970), who conclude that the Ni(100)c(2 × 2)–O pattern is due to the formation of an oxide of composition Ni_3O. The situation remains controversial with LEED workers supporting adsorption sites with two-fold or four-fold coordination. The situation is reviewed in Watson (1987). As the controversy depends upon the interpretation of two different models (a model for LEED $I(V)$ curves and a model for the oxidation process) it may not be resolved until a third kind of observation, less dependent on models, can be made on the system.

Pd(111)–CO   The work of Ertl and Koch (1972) provides a good example of the way in which several techniques can be combined so as to draw conclusions about the atomic details of an adsorption process. The adsorption of carbon monoxide upon the clean Pd(111) surface was studied by means of LEED, Auger spectroscopy, mass spectroscopy, work function measurement, and flash desorption. Flash desorption shows a single peak at about 470 K, from which it is concluded that there is a single binding state on Pd(111). The area under this peak could be used to determine the coverage of carbon monoxide and so the work function change on adsorption could be related to coverage. The work function increased on adsorption, which suggests polarization of the carbon monoxide admolecules with a negative pole towards the vacuum. The variation of work function with coverage, taken together with a knowledge of the arrival rate of carbon monoxide molecules impinging from the ambient carbon monoxide atmosphere, enables the variation of sticking coefficient S with coverage θ to be evaluated. This shows initial behaviour rather like Fig. 6.4(a) with $S=1$ until θ is about 0.2 monolayers. S then falls linearly to zero at $\theta = 0.5$ monolayers. By measuring the partial pressures of carbon monoxide and the work function at various temperatures the isosteric heat of adsorption could be estimated (isosteric means constant θ), and is found to be 1.47 eV up to $\frac{1}{3}$ of a monolayer.

The first ordered LEED pattern at room temperature is Pd(111)–($\sqrt{3} \times$

$\sqrt{3}$)R30°–CO (Fig. 6.10(a)), which occurs at $\theta = \frac{1}{3}$. There are three equivalent domain orientations of this structure, which are indicated in Ertl and Koch's proposed model of the carbon monoxide adsorption shown in Fig. 6.10(b). In order to suggest this model the work function changes described above, together with infrared spectroscopic studies on the same system by other observers, are used to conclude that the carbon is adjacent to the metal surface in a 'bridge-bonded' site indicated in Fig. 6.10(c). The axis of the carbon monoxide is normal to the surface. The size of the carbon monoxide molecule is such that there is space only for one in each unit mesh of the adlayer. Similar structure models for carbon monoxide adsorption on Pd(10), Rh(111), Ni(100), and Ir(111) have been proposed by other authors.

Surface segregation

Another way that atoms or molecules can arrive at a surface and become adsorbed is by diffusion out of the bulk. This is the process of *surface segregation*. This may occur when an impurity in a solid has a higher or lower concentration at the surface than it does in the bulk. Alternatively, it may be that a binary or more complex alloy also has a different concentration at the surface than in the bulk. In the case of a dilute binary alloy, if the minority component is more concentrated at the surface then the phenomenon is described as *solute segregation*. If the majority component is more concentrated at the surface then it is known as *solvent segregation*. This is a long established field of study, dating from about a century ago when Willard Gibbs studied the thermodynamics of segregation. It is of great importance in metallurgy, where, for instance, the embrittlement of an alloy can be caused by segregation of impurities to grain boundaries (a kind of internal surface) and the corrosion of the material can be strongly affected by segregation to a surface.

A comparison of the experimental results for segregation in dilute binary alloys with a simple theory has been described by Abraham and Brundle (1981). These authors consider that there are two important effects at work in driving the diffusion of one atomic species towards or away from the surface:

1. The sizes of the two kinds of atoms in the alloy may be different. Thus there will be a difference between the strain energy of the solid when the atoms are randomly distributed throughout it and when there is more of one kind of atom at the surface.

2. The strength of the bonds between atoms of one kind and those of the other kind will generally be different. By breaking bonds with low binding energies and forming more bonds with high binding energies the free energy of the system can be reduced. This bond-breaking provides a driving force to cause diffusion of one of the species towards the surface where the coordination numbers are lower.

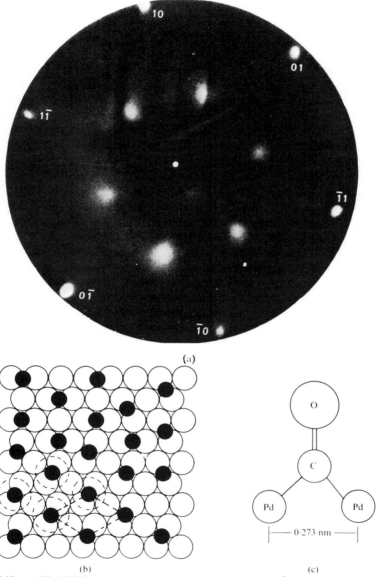

Fig. 6.10 (a) The LEED pattern at 34 eV for Pd(111)–CO. It is a $\sqrt{3}$ R30°–CO pattern. Room temperature; $\theta = \frac{1}{3}$. (b) Ertl and Koch's model to account for this LEED pattern. The open circles are the palladium surface atoms, the solid circles are carbon atoms, and the broken circles are oxygen atoms. One of the three equivalent domain orientations is indicated by its unit mesh. (c) The way the carbon monoxide molecule is thought to 'stand up' on the surface with a bridge bond between palladium atoms to the carbon. The carbon monoxide is thus bonding with twofold coordination unlike the fourfold coordination of the oxygen in Ni(100)c(2 × 2)–O shown in Fig. 6.9.

This pair of driving forces had been recognized for some time, but theories of segregation had involved using only one of the forces or a linear combination of both. Abraham and Brundle considered that it was not evident that linear elasticity theory or linear combinations of essentially macroscopic descriptions would be appropriate at an atomic level. Accordingly, they considered various solid solution systems of atoms in bulk and surface configurations and minimized the total potential energy of the system in each case. This minimization allowed the positions of the atoms around a solute atom to be relaxed. The situation is indicated in Fig. 6.11.

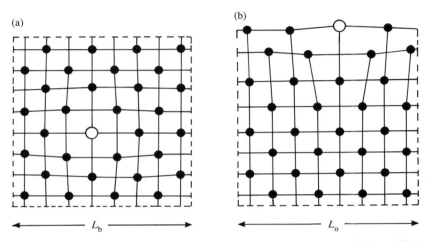

Fig. 6.11   Representations of atomic relaxation about a solute atom in (a) the bulk and (b) the surface of a matrix of solvent atoms. (By courtesy of Abraham and Brundle 1981.)

This calculation led to the idea of a plot of the ratio γ^* of the surface tensions of the pure elemental materials versus the ratio σ^* of their atomic sizes. The former reflects the contribution of the surface free energy and the latter the contribution of the lattice strain to the surface segregation. Such a plot is shown in Fig. 6.12. In the shaded area of the plot the calculations predict that the solvent atoms will segregate to the surface and elsewhere the solute atoms are segregated to the surface. If this prediction is correct and if the ratios γ^* and σ^* can be established from measurements then this plot provides a useful means of ascertaining the expected segregant. The atomic size ratio can be deduced from the known crystallography of the materials. The surface tension ratio can be estimated from experiments to measure the surface tensions in the liquid state of the materials. Therefore it is possible to estimate the position of any particular dilute alloy in the plot of Fig. 6.12.

Experiments to measure surface segregation and test the plot in Fig. 6.12 can be quite difficult. Problems can arise from:

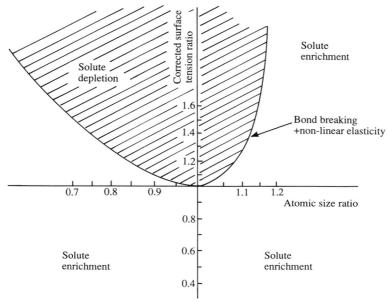

Fig. 6.12 The structure of the surface tension ratio–atomic size ratio plot for predicting surface segregation.

1. The segregation of non-metallic bulk impurities which can alter the observed surface concentrations of the elements of the alloy.

2. The need to observe the equilibrium surface conditions at elevated temperatures (often above 1000 K) in UHV. This means that XPS or AES measurements should be made at the temperature for which equilibrium segregation is being observed. This difficulty can be avoided by establishing equilibrium and cooling the sample rapidly to make the spectroscopic observations. This 'quenches' in the state appropriate to the higher temperature while allowing a simpler measurement at the lower temperature.

3. The extreme time-scales for the experiments can present some difficulty. At the lower temperatures the diffusion times to reach equilibrium surface concentrations can be very many hours and stringent UHV conditions may not be sustainable. At the higher temperatures segregation can be so rapid that the time required to quench the sample to the measurement temperature is long enough for changes to have occurred.

4. The process of segregation is usually anisotropic—it occurs at different rates and to different equilibrium levels on different crystal surfaces. This means that a characterization of a particular material involves measurements on many different orientations of single-crystal samples if a technique with a

large sampling area is used (e.g. conventional AES or XPS). If a microscopic technique is used (such as scanning Auger microscopy) then a polycrystalline sample with many different exposed crystal faces can be studied, but the experimental times are still very long.

Nevertheless, a great deal of careful experimental work has been done in the study of segregation, and Abraham and Brundle have compared the results of this work with their model predictions. The results for 45 bimetallic dilute alloys were compared with the plot in Fig. 6.12 and the agreement found to be very good. Therefore this appears to be a good quide to the kind of segregation to be expected.

Epitaxial processes

The gas–solid interactions described above may be the first stages in the growth of an oriented single-crystal film of one material upon a single-crystal substrate of another. This process is called *epitaxial growth*. An example is the formation of an oriented layer of nickel monoxide upon nickel—an epitaxial process that can proceed in the sequences

$$Ni(100) + O_2 \xrightarrow{\text{R.T.}} \underset{\text{adlayer}}{Ni(100)p(1 \times 1)\text{–}O} + \underset{\text{substrate}}{Ni(100)}$$

$$Ni(100)p(1 \times 1) + O_2 \xrightarrow{\text{R.T.}} \underset{\text{adlayer}}{Ni(100)c(2 \times 2)\text{–}O} + \underset{\text{substrate}}{Ni(100)}$$

$$Ni(100)c(2 \times 2)\text{–}O + O_2 \xrightarrow{400^\circ C} \underset{\substack{\text{epitaxial} \\ \text{film}}}{NiO(100)} + \underset{\text{substrate}}{Ni(100)}$$

Examples of epitaxial processes are found not only in such gas–solid interactions but also in the deposition of many materials upon single crystals of very many others. The subject is reviewed in the two volumes edited by Matthews (1975).

Just as the situation with respect to adsorption is complex so it is with respect to epitaxial growth. There is no theoretical fabric which enables predictions to be made as to whether or not a particular material will grow epitaxially upon another, and, if so, in what orientation. What process occurs depends at least upon the adatom–adatom, substrate-atom–substrate-atom, and substrate-atom–adatom bond strengths, the incident flux, the substrate temperature, and the surface diffusion coefficient of the adatom. Very often, insufficient data are available about all these parameters for the observations of epitaxial processes to be arranged in meaningful patterns.

Most studies of epitaxial processes have used the techniques of electron microscopy (Pashley 1970), which normally involve removing the epitaxial film from its substrate and transferring it, through the air, from the deposition system to the electron microscope. Recently, *in situ* observations in special UHV electron microscopes have been possible and, in addition, the techniques of LEED and Auger electron spectroscopy have been used. The impact of these surface techniques upon the understanding of epitaxial processes is reviewed by Bauer and Poppa (1972).

One way of classifying processes of epitaxial growth is by the mode of growth, as indicated in Fig. 6.13. After specifying the mode of growth it is necessary also to define the orientation of the deposited layer with respect to the substrate. If the adlayer is only of the order of a monolayer thick, then the notation used so far in this book is adequate to describe the deposit orientation. Thicker oriented films are described by specifying the deposit plane, which is parallel to the substrate surface plane, and also a direction in the deposit surface plane which is parallel to a direction in the substrate surface plane. Examples are given later.

Provided that the deposit does not alloy with the substrate, or that no gross changes in the substrate surface structure occur during deposition (e.g. dissociation of a substrate due to electron bombardment or chemical reaction between substrate and deposit) then the modes of Fig. 6.13 can be understood in terms of the relative surface energies of the deposit and substrate materials. The surface energy is simply the excess internal energy of the solid–vacuum system over that of an imaginary system with two homogeneous phases separated by an ideally discontinuous change at a mathematical surface between them (e.g. Blakely 1973). The surface energy is different from the bulk energy because of the broken bonds at the surface and the possible relaxation or rumpling (Chapter 1). For nucleation to occur (Fig. 6.13(a)) the surface energy of the deposit material is high compared to that of the substrate. For monolayers to form at the substrate surface (Fig. 6.13(b) or (c)) the deposit must have a lower surface energy than the substrate. Bauer and Poppa (1972) distinguish monolayer growth (Fig. 6.13(b)) from nucleation after monolayer growth (Fig. 6.13(c)) by recognizing that, if the deposit is strained so as to 'match' deposit lattice spacings to substrate lattice spacings, then the growth mode will depend upon the relative sizes of the deposit strain energy and the deposit surface energy. If the strain energy in the deposit is low compared with its surface energy, then monolayer growth is expected. If, however, the deposit strain energy is high it may become defective in some way after a monolayer is formed. This defectiveness may show up as dislocations in an otherwise flat layer, or the material may nucleate on top of the first monolayer, as in Fig. 6.13(c).

In addition to these different growth modes, different relative orientations of deposit and substrate can occur. If the deposit grows in parallel orientation with the substrate surface, the deposit atoms building up as if to continue the atomic structure of the substrate, then the process is said to be

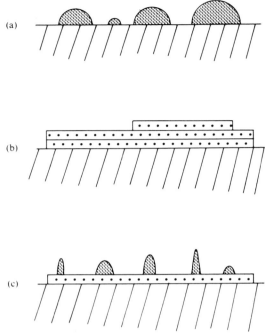

Fig. 6.13 Modes of growth in an epitaxial process. (a) The deposit nucleates on the substrate surface either randomly, as adatoms meet by accident and form stable clusters, or at special defect or impurity sites in the substrate surface (see Fig. 5.4). Nuclei then grow by addition of adatoms directly from the impinging vapour or from surface diffusion. The nuclei may rotate on the substrate. Coalescence to a film occurs later and orientation changes can happen at this stage of growth. This is known as the *Volmer–Weber mode*. (b) The deposit grows in monatomic adlayers first upon the substrate and subsequently upon itself. This is known as the *Frank–van de Merwe mode* or simply the *monolayer growth mode*. (c) The first atoms arriving at the substrate form an atomic monolayer and subsequent atoms nucleate to form islands on top of the monolayer. This is the *Stranski–Krastanov mode*.

pseudomorphic. If a film of metal is grown upon a single-crystal surface of another metal which has a lattice spacing close to that of the deposit, then pseudomorphic epitaxy is often observed. An example of such a system is nickel on copper. Nickel and copper differ in bulk lattice spacings by only 2.5 per cent, and nickel is found to grow pseudomorphically upon Cu(100) and Cu(111) surfaces with the mode of Fig. 6.13(b). As the thickness of nickel is increased it stores more and more strain energy, until a critical thickness is reached where the strain causes dislocations to be introduced and the nickel can relax towards its bulk lattice constant. This case is discussed, for instance, by Joyner and Somorjai (1973). If strict pseudomorphism were occurring here, the depositing nickel atoms would be located in copper 'sites' as indicated in an exaggerated way in Fig. 6.14. It is possible to use LEED $I(V)$ studies (Chapter

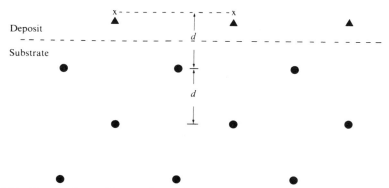

Fig. 6.14 The pseudomorphic growth problem. In strict pseudomorphism depositing, atoms (e.g. nickel) would be located at substrate atom sites marked X (e.g. copper). If the nickel–copper spacing d is not strictly pseudomorphic the deposit atoms may be located with a nickel–nickel spacing or some intermediate spacing (marked ▲) appropriate perhaps to an interfacial alloy.

3) to determine the vertical spacing of the nickel adatoms; this has been attempted and the nickel atoms are indeed in the copper sites for the first monolayer of nickel.

A similar example which has been studied by almost every technique described in this book is W(110)–Ag. Here, instead of strict pseudomorphism, the first monolayer is an Ag(111) layer which is strained to fit the tungsten atomic spacings along the W[001] direction. As more silver is deposited further monolayers grow, but with different periodicities, until finally an epitaxial silver layer is formed with Ag {111} [112] parallel to W(110)[112]. This case is described by Bauer and Poppa (1972).

On the other hand, a deposit may grow as a single crystal with the bulk lattice spacing from the earliest observable stages. Such a growth process is observed in the much studied alkali halide/metal systems. These systems were of great interest because the epitaxial film could be dissolved off its alkali halide substrate and floated on to a metallic support grid for study by transmission electron microscopy. All these systems nucleate as in Fig. 6.13(a), with islands usually in a (100) parallel epitaxial orientation or with a (111) plane parallel to the (100) cleavage face of the alkali halide. An example of the parallel orientation in KCl(100)–Ag is shown in Fig. 6.15. In these cases the mismatch between the lattice constants of substrate and deposit can be very large (sometimes as much as 30 per cent) and so the strain energy that would be associated with pseudomorphic growth renders it unsupportable. It appears that (100) oriented islands are formed by clusters of four adatoms strongly bonded together and less strongly bound to the substrate. The orientation of this cluster is determined by its interaction with the substrate surface—as indicated in Fig. 6.16. In the case of metals upon alkali halides, the

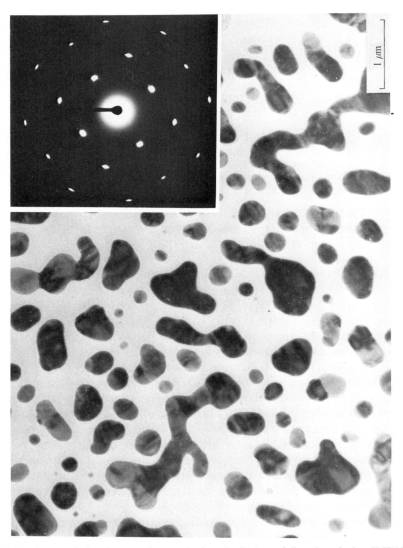

Fig. 6.15 A transmission electron micrograph of nominally 4 nm of silver deposited on KCl(100) at 320 K. The silver islands are supported on an amorphous film of carbon about 20 nm thick added after the epitaxial deposition. The potassium chloride substrate was electron bombarded during deposition—a process which improves the (100) epitaxy of the deposit. *Inset*: Transmission diffraction pattern of the same region as the micrograph, obtained with electrons of energy 100 keV. (From Lord and Prutton 1974.)

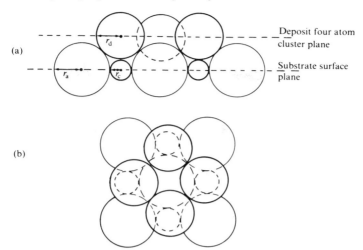

(a)

(b)

Fig. 6.16 The four-atom cluster for epitaxial nucleation with (100)[001] deposit parallel to (100) [100] substrate. (Due to Vermaak and Henning 1970.) (a) Section containing [001] in substrate and deposit; (b) plan looking down on a cluster in the (100) plane. Such clusters are possible if

$$r_d < \frac{r_a^2 - r_c^2 + d^2}{(2r_c + 2^{1/2}d - r_a)}$$

$$d = r_a + r_c$$

anisotropic character of this interaction (which must be the feature determining the orientation of the cluster) is not very strong and not only can several orientations occur, but the balance between them can be disturbed by the process of coalescence. The orientation of the first continuous film is then not very clearly related to the dominating orientation amongst the initial clusters. The subject is complex and the interested reader is referred to reviews by Bauer and Poppa (1972), Lord and Prutton (1974), and the two volumes edited by Matthews (1975).

In the example of a four-atom cluster taking up a particular orientation upon an alkali halide (100) surface, two important parameters in the theoretical description of the problem will be the adatom–substrate interaction potential and the adatom–adatom interaction potential. Little is known of the former and very little of the latter in spite of its importance in determining whether or not odered overlayers will form. The experimental difficulty in observing such interactions is considerable, but has been possible using the atomic resolving power of FIM (Chapter 3). If the adatoms are sufficiently strongly bound to the surface of the tip in a field-ion microscope and yet still mobile enough to diffuse around and come into positions of minimum energy with respect to each other, then adatom–adatom interactions can be observed. In some particularly elegant work, Basset (1973) has been able to observe stable two-atom clusters—*dimers*—of Ta_2, W_2, Ir_2, Pt_2, and WRe upon W(110) surfaces. Iridium is particularly striking in that it can form long parallel adatom chains (Fig. 6.17) which forms about 0.15 nm apart.

If the adatom–substrate interaction is strong but the adatom-adatom interaction is weak, then the growth mechanism of Fig. 6.13(c)—monolayer followed by nucleation—can occur. Such a mechanism is observed for alkali metals deposited upon tunsten (Mayer 1971).

The wealth of examples of epitaxial growth is as great as the paucity of explanations for their occurrence. It is to be hoped that clearer patterns will emerge as the great variety of parameters required to describe these processes are measured using the kind of techniques described in the chapters above. The techniques of STM are likely to be very valuable here because of their ability to observe the adsorption sites, the coalescence, and the movement of atoms.

Molecular beam epitaxy (MBE)

Molecular beam epitaxy is a sophisticated form of vacuum evaporation in which atomic or molecular beams are directed at a carefully prepared single-crystal substrate in a very clean UHV system. It has grown into prominence because these clean and controlled deposition conditions have been found to result in a variety of metal/metal and semiconductor/semiconductor systems which have extraordinary electrical or magnetic properties. The archetypal system is a multilayer structure of GaAs and GaAlAs in which a layer of GaAs is sandwiched between thicker layers of GaAlAs. If the thickness of the GaAs is less than an electron's mean free path, then it behaves as a two-dimensional (2D) well containing the free electron gas. It is a *quantum well*. Such layers can show extraordinarily high electron mobilities and unusual effects like the *quantum Hall effect*. For these reasons there has been considerable interest in the growth mechanisms and the physics and applications of these structures. The design and understanding of MBE systems has been reviewed by Joyce (1985) in a special issue of *Reports on Progress in Physics*, in which quantum well devices are discussed by Board and the physics of quantum well devices are discussed by Kelly.

A typical MBE system is outlined in Fig. 6.18 which is due to Joyce. The UHV arrangements for such apparatus are usually quite complex in order to ensure very low levels of contamination of the films being deposited with impurity atoms from the ambient atmosphere, the vapour sources and their crucibles and heating arrangements and any supplementary monitoring or measuring apparatus. Thus, they are likely to have ion pumps and cryopanels for achieving and maintaining the vacuum and air-lock systems for the introduction and removal of substrates without exposing the vacuum chamber to atmospheric pressure.

The structures of multiple quantum wells grown in MBE systems can be

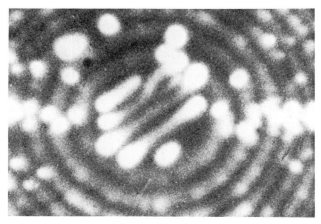

Fig. 6.17 Parallel adatom chains with a separation of 0.15 nm in a deposit of (80 ± 5) iridium atoms formed on a W(110) plane on a tip in an FIM. Iridium vapour was deposited at 78 K and then the tip was heated to 380 K. (After Bassett 1973.)

beautifully revealed by cutting sections through them with chemical etchants. Thin layers cut perpendicular to the wells can then be examined in the transmission electron microscope (TEM) and the layer structure revealed. This is known as *cross-sectional TEM*. An example of such TEM results is shown in Fig. 6.19. With careful inspection the periodic contrast due to the crystal lattice potential can be seen both in the GaAlAs buffer layers and in the darker GaAs quantum well. Further, the registration of the layers can be seen at the interfaces and the transition from one material to the other can be seen to be very sharp—it occurs in a single atom spacing.

Summary

The problems described in this chapter are amongst the most challenging and the least resolved in surface physics. The adsorption of single atoms upon single-crystal surfaces is the first step in many kinds of processes and yet it is relatively poorly understood. It requires knowledge of the way electron exchange can occur with substrate, to what extent the substrate atoms relax about the adatom, how lattice vibrations affect the adsorption of adatoms, and a precise knowledge of the position in which the adatom is located. Measurements of work function and energy distribution of emitted electrons and observations using LEED, Auger electron emission, STM, UPS, XPS, and RHEED can help to provide this information, but the power of combinations of these techniques has rarely been applied to one carefully characterized system. The success attendant upon this approach, in the few

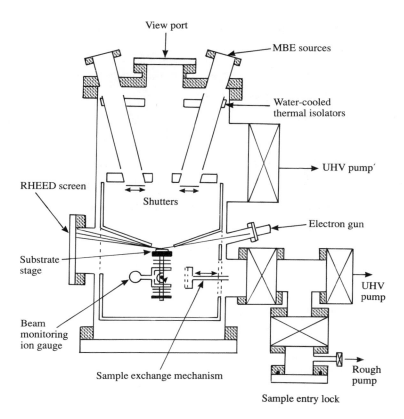

Group III source (Ga, Al)

Fig. 6.18   Diagram of an MBE system. The atomic or molecular beams are provided by Knudsen cells which are evaporation sources with a carefully controlled temperature inside a cell with a very small aperture through which the beam can escape towards the substrate. The aperture is sufficiently small that the vapour pressure inside the cell is maintained at its equilibrium value, so that the small escape rate due to effusion is determined solely by the temperature. The deposition rate at the substrate is then controlled with the source temperature and the geometry of the arrangement. The MBE system usually contains a variety of analytical tools. RHEED (Chapter 3) is very common and is used both to monitor the crystallographic order in the substrate and the deposit as fabrication proceeds and to determine the deposit thicknesses by exploiting the RHEED intensity oscillations also described in Chapter 3. Auger electron spectroscopy is incorporated for monitoring the film composition and characterizing the surface cleanliness (Chapter 2). (By courtesy of Prof. B. A. Joyce.)

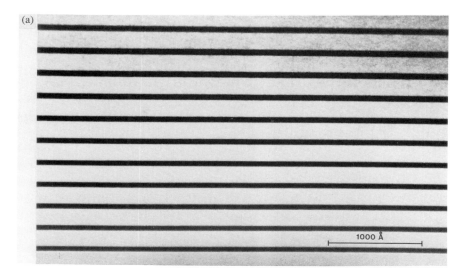

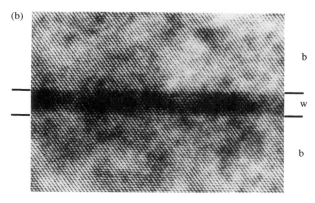

Fig. 6.19 (a) (110) cross-sectional TEM micrograph of a multiple quantum well structure. The GaAs wells are 5.5 nm thick and the GaAlAs barriers are 17.5 nm thick. The image is obtained using dark field illumination in which the (002) diffracted beam is collected through the objective aperture of the electron microscope and used by subsequent lenses to form an image. (b) A high-resolution (110) cross-sectional TEM micrograph of a 2.7 nm thick GaAs quantum well (thickness indicated as w) and GaAlAs barriers. (By courtesy of Prof. B. A. Joyce, Dr J. Gowers, and the Philips Research Laboratories, Salford, England.)

cases in which it has been used, encourages further work. The case of carbon monoxide adsorption on f.c.c. metal faces is an example of such an approach.

The same conclusions can be drawn about the second difficult problem described in this chapter—epitaxial growth. Again, a good example is

W(110)–Ag. An understanding of the interplay of the important parameters—the adatom–substrate and adatom–adatom interaction potentials—can be provided by combinations of the kinds of techniques described here, but it is unlikely to be revealed by only one of them.

References

Preface

Blakely, J. M. (1973). *Introduction to the properties of crystal surfaces.* Pergamon Press, Oxford.
Briggs, D. and Seah, M. P. (1990). *Practical surface analysis. Volume 1: Auger and X-ray photoelectron spectroscopy.* Wiley, Chichester.
Briggs, D. and Seah, M. P. (1992). *Practical surface analysis: Volume 2: Ion and neutral spectroscopy.* Wiley, Chichester.
Kittel, C. (1986). *Introduction to solid state physics.* Wiley, New York.
Rosenberg, H. M. (1975). *The solid state.* Clarendon Press, Oxford.
Somorjai, G. A. (1972). *Principles of surface chemistry.* Prentice-Hall, Englewood Cliffs NJ.
Woodruff, D. P. and Delchar, T. A. (1988). *Modern techniques of surface science.* Cambridge University Press, Cambridge.
Zangwill, A. (1988). *Physics at surfaces.* Cambridge University Press, Cambridge.

Chapter 1

Bond, G. C. (1974). *Heterogeneous catalysis: principles and applications.* Clarendon Press, Oxford.
Chambers, A., Fitch R. K., and Halliday B. S. (1989). *Basic vacuum technology.* Adam Hilger, Bristol.
Kittel, C. (1986). *Introduction to solid state physics.* Wiley, New York.
Redhead, P. A. (1968). *The physical basis of ultra-high vacua.* Chapman & Hall, London.
Rosenberg, H. M. (1975). *The solid state.* Clarendon Press, Oxford.

Chapter 2

Bassett, P. J., Gallon, T. E., Matthew, J. A. D., and Prutton, M. (1972). *Surf. Sci.,* **33,** 213.
Bearden, J. A. and Burr, A. F. (1967). *Rev. Mod. Phys.,* **39,** 125.
Benninghoven, A. (1973). *Surf. Sci.,* **35,** 427.
Bishop, H. E. and Riviere, J. C. (1969). *J. Appl. Phys.,* **40,** 1740.
Briggs, D. and Seah, M. P. (1990). *Practical surface analysis. Volume 1: Auger and X-ray photoelectron spectroscopy.* Wiley, Chichester.
Briggs, D. and Seah, M. P. (1992). *Practical surface analysis. Volume 2: Ion and neutral spectroscopy.* Wiley, Chichester.
Brundle, C. R. and Baker, A. D. (1981). *Electron spectroscopy: theory, techniques and applications.* Academic Press, London.
Carlson, T. A. (1975). *Photoelectron and auger spectroscopy.* Plenum Press, New York.
Cerezo, A., Godfrey, T. J., Grosvenor, C. R. M., Hetherington, M. G., Hoyle, R. M., Jakubovics, J. P., Liddle, J. A., Smith, G. D. W., and Worrall, G. M. (1989). *J. Microsc.,* **154,** 215.
Cini, M., (1978). *Phys. Rev.,* **B17,** 2788.

Fadley, C. S. and Shirley, D. A. (1970). *J. Res. Natn. Bur. Stand.*, **74A**, 543.

Gelius, U., Wannberg, B., Baltzer, P., Fellner-Feldegg, H., Carlsson, G., Johansson, C.-G., Larsson, J., Munger, P., and Vegerfors, G. (1990). *J. Electron Spectrosc. Rel. Phen.*, **52**, 747.

Jackson, D. J., Chambers, A., and Gallon, T. E. (1973). *Surf. Sci.*, **36**, 381.

Kittel, C. (1986). *Introduction to solid state physics*. Wiley, New York.

Kuhn, H. G. (1969). *Atomic spectra*. Longman Green, London.

Madden, H. H. (1981). *J. Vac. Sci. Technol.*, **18**, 677.

Maissel, L. I. and Chang, R. (1970). *Handbook of thin film technology*. McGraw-Hill, New York.

Muller, E. W. and Tsong, T. T. (1969). *Field ion microscopy*. Elsevier, New York.

Niblack, W. (1986). *Digital image processing*. Prentice-Hall, Englewood Cliffs NJ.

Redhead, P. A. (1968). *The physical basis of ultra-high vacua*. Chapman & Hall, London.

Sevier, K. D. (1972). *Low energy electron spectrometry*. Wiley-Interscience, New York.

Siegbahn, K., Nordling, C., Fahlman, A., Nordberg, R., Hamrin, K., Hedman, J., Johannson, G., Belgmark, T., Karlsson, S. E., Lindgren, I., and Lindberg, B. (1967). *ESCA—Atomic, molecular and solid state structure studied by means of electron spectroscopy*. Almquist and Wiksell, Uppsala.

Smialek, J. L. and Browning R. (1986). In *Proc. symp. high temperature materials chemistry*, p. 258. The Electrochemical Society, Pennington NJ.

Spicer, W. E. (1970). *J. Res. Natn. Bur. Stand.*, **74A**, 397.

Tanuma, S., Powell, C. J., and Penn, D. R. (1990). *J. Electron Spectrosc. Rel. Phen.*, **52**, 285.

Wagner, C. D. and Joshi, A. (1988). *J. Electron Spectrosc. Rel. Phen.*, **47**, 283.

Walker, C. G. H., Peacock, D. C., Prutton, M., and El Gomati, M. M. (1988). *Surf. Interface Anal.*, **11**, 266.

Weightman, P. (1982). *Rep. Progr. Phys.*, **45**, 753.

Wells, O. C., Boyde, A., Lifshin, E., and Rezanowich, A. (1974). *Scanning electron microscopy*. McGraw-Hill, New York.

Chapter 3

Andersen, J. N., Nielsen, H. B., Petersen, L., and Adams, D. L. (1984). *J. Phys. C: Solid State Phys.*, **17**, 173.

Bauer, E. (1962). *Proc. 5th International Congress on Electron Microscopy*. (ed. S. S. Breese Jr.) Academic Press, New York. 11.

Bauer, E. (1985). *Ultramicroscopy*, **17**, 51.

Bauer, E. and Telieps, W. (1988). In *Surface and interface characterization by electron optical methods*, (eds A. Howie and U. Valdre). Plenum Press, New York.

Binnig, G. and Rohrer, H. (1983). *Phys. Rev. Lett.*, **50**, 120.

Chadi, D. J. (1978). *Phys. Rev.*, **B18**, 1800.

Cowell, P. G. and de Carvalho, P. G. (1988). *J. Phys. C: Solid State Phys.*, **21**, 2983.

Davisson, C. J. and Germer, L. H. (1927). *Phys. Rev.*, **30**, 705.

Fadley, C. S., Kono, S., Petersson, L. G., Goldberg, S. M., Hall, N. F. T., Lloyd, J. T., and Hussain, Z. (1979). *Surf. Sci.*, **89**, 52.

Fiedenhans'l, R. (1989). *Surf. Sci. Rep.*, **10**, 105.

Gervais, A., Stern, R. M., and Menes, M. (1968). *Acta Cryst.*, **24**, 191.

Guntherödt. H.-J. and Wiesendanger, R. (1992). *Scanning Tunneling Microscopy I*. Springer, Berlin.

Heidenreich, R. D. (1964). *Fundamentals of transmission electron microscopy*. Wiley, New York.

Henzler, M. (1984). *Applied Physics*, **A34**, 205.

Jona, F., Strozier, J. A., and Wang, W. S. (1982). *Rep. Progr, Phys.*, **45**, 527.

Joyce, B. A., Neave, J. H., Zhang, J., Dobson, P. J., Dawson, P., Moore, K. J., and Foxon, C. T. (1986). In *Thin film growth techniques for low-dimensional structures* (eds R. F. C. Farrow, S. S. P. Parkin, P. J. Dobson, J. H. Neave, and A. S. Arrott). Plenum Press, New York.

Kuk, Y. (1992). In *Scanning tunneling microscopy I*. (ed. H.-J. Güntherodt and R. Wiesendanger), pp. 17–38. Springer-Verlag, Heidelberg.

Leemput, van de L. E. C. and Lekkerkerker, H. N. W. (1992). *Rep. Progr. Phys.*, **55**, 1165.

Maksym, P. A. and Beeby, J. L. (1984). *Surf. Sci.*, **140**, 1.

Masud, M., Kinniburgh, C. G., and Pendry, J. B. (1977). *J. Phys. C: Solid State Phys.*, **10**, 1.

Muller, E. W. (1951). *Z. Phys.*, **131**, 136.

Muller, E. W. (1965). *Science*, **149**, 591.

Muller, E. W. (1970). In *Modern diffraction and imaging techniques in materials science* (eds S. Amelinckx, R. Gevers, G. Remault, and J. Van Landuyt). North-Holland, Amsterdam.

Muller, E. W., Panitz, J. A., and McLane, S. B. (1968). *Rev. Sci. Instrum.*, **39**, 83.

Norton, P. R., Davies, J. A., Jackson, D. P., and Matsunami, N. (1979). *Surf. Sci.*, **85**, 269.

Pendry, J. B. (1974). *Low energy electron diffraction*. Wiley-Interscience, New York.

Rosenberg, H. M. (1974). *The solid state*. Clarendon Press, Oxford.

Rous, P. J., Pendry, J. B., Saldin, D. K., Heinz, K., Muller, K., and Bickel, N. (1986). *Phys. Rev. Lett.*, **57**, 2951.

Rous, P. J. (1990). In *The structure of surfaces III* (eds. S. Y. Tong, M. A. Van Hove, K. Takayanagi, and X. D. Xie), p. 118. Springer, Berlin.

Stohr, J. (1979). *J. Vac. Sci. Technol.*, **16**, 37.

Van Hove, M. A. and Tong, S. Y. (1979). *Surface crystallography by LEED*. Springer, Berlin.

Van Hove, M. A., Weinberg, W. H., and Chan, C.-M. (1986). *Low energy electron diffraction*. Springer, Berlin.

Van Hove, M. A. (1991). In *Structure of solids* (ed V. Gerold). Verlag Chemie, Weinheim.

Warburton, D. R., Thornton, G., Norman, D., Richardson, C. H., McGrath, R., and Sette, F. (1987). *Surf. Sci.*, **189/190**, 495.

Watson, P. R. (1987). *J. Phys. Chem. Ref. Data*, **16**(4), 953.

Wille, K. (1991). *Rep. Progr. Phys.*, **54**, 1069.

Wood, E. A. (1964). *J. Appl. Phys.*, **35**, 1306.

Woodruff, D. P. and Delchar, T. A. (1989). *Modern techniques of surface science*. Cambridge University Press, Cambridge.

Woolfson, M. M. (1971). *An introduction to X-ray crystallography*. Pergamon Press, Oxford.

Wormald, J. (1973). *Diffraction methods*. Clarendon Press, Oxford.

Chapter 4

Adams, D. L. and Germer, L. H. (1971). *Surf. Sci.* **27**, 21.

Allyn, C. L., Gustafsson, T., and Plummer, E. W. (1977). *Phys. Rev. Lett.*, **47**, 127.

Arlinghaus, F. J., Gay, J. G., and Smith, J. R. (1980). *Phys. Rev.*, **B21**, 2055.

Beshara, N. M., Buckman, A. B., and Hall, A. C. (1969). In *Proc. Symp. Recent Developments in Ellipsometry*. North-Holland, Amsterdam.

Binnig, G., Rohrer, H., Gerger, C., and Wiebel, E. (1983). *Phys. Rev. Lett.*, **50**, 120.

Bradshaw, A. M., Cederbaum, L. S., and Domcke, W. (1975). *Ultraviolet photoelectron spectroscopy of gases adsorbed on metal surfaces. Struct. Bonding*, **24**, 133.

Chadi, D. J. (1984). *Phys. Rev.*, **B30**, 4470.

Eberhardt, W. and Himpsel, F. J. (1979). *Phys. Rev. Lett.*, **42**, 1375.

Farnsworth, H. E., Shlier, R. E., and Dillon, J. A. (1959). *J. Phys. Chem. Solids*, **8**, 116.

Fowler, R. (1933). *Phys. Rev.*, **38**, 45.

Gadzuk, J. W. (1972). *J. Vac. Sci. Technol.*, **9**, 591.

Hamers, R. J., Tromp, R. M., and Demuth, J. E. (1986). *Phys. Rev. Lett.*, **56**, 1972.

Hamers, R. J. in Guntherödt, H.-J., and Wiesendanger, R. (eds) (1992). *Scanning tunneling microscopy I*. Springer, Berlin.

Haneman, D. (1987). *Rep. Progr. Phys.*, **50**, 1045.

Heavens, O. S. (1964). *Measurement of optical constants of thin film*. In *Physics of thin films*, Vol. 2 (eds. G. Haas and R. E. Thun). Academic Press, New York.

Heimann, P., Hermanson, J., Miosga, H., and Neddermayer, H. (1979). *Phys. Rev.*, **B20**, 3059.

Himpsel, F. J. and Fauster, T. (1984). *J. Vac. Sci. Technol.*, **A2**, 815.

Ihm, J., Cohen, M. L., and Chadi, D. J. (1980). *Phys. Rev.*, **B21**, 4592.

Inglesfield, J. E. (1982). *Rep. Progr. Phys.*, **45**, 223.

Kerker, G. P., Ho, K. M., and Cohen, M. L. (1978). *Phys. Rev.*, **B18**, 5473.

Kittel, C. (1986). *Introduction to solid state physics*. Wiley, New York.

Lang, N. D. and Kohn, W. (1970). *Phys. Rev.*, **B1**, 4555.

Larsen, P. K., Chiang, S., and Smith, N. V. (1977). *Phys. Rev.*, **B15**, 3200.

Lunsford, J. H. (1972). *Adv. Catalysis*, **32**, 265.

Malus, E. L. (1808). *Nov. Bull. Soc. Philomath.*, **1**, 266.

McKelvey, J. P. (1966). *Solid state and semiconductor physics*. Harper & Row, New York.

Muller, E. W. (1970). *Modern diffraction and imaging techniques in materials science* (eds. S. Amelinckx, R. Gevers, G. Remault, and J. Van Landuyt). North-Holland, Amsterdam.

Porteus, J. O. and Faith, W. N. (1973). *Phys. Rev.*, **B8**, 491.

Riviere, J. C. (1969). *Solid State Surf. Sci.*, **1**, 179.

Robinson, I. K., Waskiewicz, W. K., Foss, P. H., Stark, J. B., and Bennet, P. A. (1986). *Phys. Rev.*, **B33**, 7013.

Rosenberg, H. M. (1974). *The solid state*. Clarendon Press, Oxford.

Rowe, J. E. and Ibach, H. (1973). *Phys. Rev. Lett.*, **31**, 102.

Somorjai, G. A. (1972). *Principles of surface chemistry*. Prentice-Hall, Englewood Cliffs NJ.

Takayanagi, K., Tanishiro, Y., Takahashi, M., Motoyoshi, H., and Yagi, K. (1984). *Electron Microsc.*, **2**, 285.

Takayanagi, K., Tanishiro, Y., Takahashi, M., and Takahashi, S. (1985). *J. Vac. Sci. Technol.*, **A3**, 1502.

Tong, S. Y., Huang, H., Wei, C. M., Packard, W. E., Men, F. K., Glander, G., and Webb, M. B. (1988). *J. Vac. Sci. Technol.*, **A6**, 615.

van der Veen, J. F., Himpsel, F. J., and Eastman, D. E. (1980). *Phys. Rev. Lett.*, **44**, 189.

Vrakking, J. J. and Meyer, F. (1971). *Appl. Phys. Lett.*, **18**, 226.

Wang, C. S., Freeman, A. J., Krakuer, H. K., and Posternak, M. (1981*a*). *Phys. Rev.*, **B23**, 1685.

Wang, C. S., Freeman, A. J., and Krakauer, H. K. (1981*b*). *Phys. Rev.*, **B24**, 3092.

Wert, C. A. and Thomson, R. M. (1970). *Physics of solids*, 2nd edn. McGraw-Hill, New York.

Williams, R. H. and McGovern, I. T. (1984). In *The chemical physics of solid surfaces and heterogenous catalysis*. (ed. D. A. King and D. P. Woodruff), vol. 3, pp. 267–309. Elsevier, Amsterdam.

Winick, H. and Doniach, S. (1980). *Synchrotron radiation research*. Plenum Press, New York.

Yamaguchi, T. (1985). *Phys. Rev.*, **B32**, 2356.

Chapter 5

Amberg, C. H. (1967). In *The solid–gas interface* (ed. E. A. Flood), vol. 2, p. 869. Arnold, London.

Bassett, D. W. (1973). In *Surface and defect properties of solids* (eds. M. W. Roberts and J. M. Thomas), vol. 2. The Chemical Society, London.

Bassett, D. W. and Parsley, M. J. (1970). *J. Phys. D: Appl. Phys.*, **3**, 707.

Blakely, J. M. (1973). *Introduction to the properties of crystal surfaces*. Pergamon Press, Oxford.

Erlich, G. (1968). *Atomistics of metal surfaces*. In. *Surface phenomena of metals*. Society of Chemical Industry, Pittsburgh, U.S.A.

Goodman, R. M. and Somorjai, G. A. (1970). *J. Chem. Phys.*, **52**, 6325.

Henrion, J. and Rhead, G. E. (1972). *Surf. Sci.*, **29**, 20.

Ibach, H. (1972). *Proc. 1st International Conference on Solid Surfaces*, p. 713. American Vacuum Society, New York.

Ibach, H. (1977). *Proc. 7th International Congress and 3rd International Conference on Solid Surfaces* (Vienna, 1977), p. 743.

Kaplan, R. and Somorjai, G. A. (1971). *Solid State Commun.*, **9**, 505.

Kittel, C. (1986). *Introduction to solid state physics*. Wiley, New York.

MacRae, A. U. (1964). *Surf. Sci.*, **2**, 52.

Rocca, M., Ibach, H., Lehwald, S., and Rahman, T. S. (1986). In *Structure and dynamics of surfaces I* (eds. W. Schommers and P. von Blanckenhagen). Springer, Berlin.

Rosenberg, H. M. (1974). *The solid state*. Clarendon Press, Oxford.

Tong, S. Y., Rhodin, T. N., and Ignatiev, A. (1973). *Phys. Rev.*, **B8**, 906.

Pluis, B., Van de Gon, A. W. D., Van der Veen, J. F., and Reimersma, A. J. (1990). *Surf. Sci.*, **239**, 265.

Willis, R. F., Lucas, A. A., and Mahan, G. D. (1983). In *The chemical physics of solid surfaces and heterogeneous catalysis* (eds. D. A. King and D. P. Woodruff). Elsevier, Amsterdam.

Woodruff, D. P. and Delchar, T. A. (1988). *Modern techniques of surface science*. Cambridge University Press, Cambridge.

Chapter 6

Abraham, F. F. and Brundle, C. R. (1981). *J. Vac. Sci. Technol.*, **18**, 506.

Anderson, P. W. (1961). *Phys. Rev.*, **124**, 41.

Bassett, D. W. (1973). In *Surfaces and defect properties of solids* (eds. M. W. Roberts and J. M. Thomas), vol. 2, p. 34. The Chemical Society, London.

Bauer, E. and Poppa, H. (1972). *Thin Solid Films*, **12**, 167.

Benedek, G. and Valbusa, U. (1982). *Dynamics of gas–surface interaction*. Springer, Berlin.

Blakely, J. M. (1973). *Introduction to the properties of crystal surfaces*. Pergamon Press, Oxford.

Board, K. (1985). *Rep. Progr. Phys.*, **48**, 1595.

Bond, C. G. (1974). *Heterogeneous catalysis, principles and applications*. Clarendon Press, Oxford.

Brennan, D. and Graham, M. J. (1966). *Discuss. Faraday Soc.*, **41**, 95.

Briggs, D. and Seah, M. P. (1990). *Practical surface analysis, Volume 1: Auger and X-ray photoelectron spectroscopy*. Wiley, Chichester.

Carroll, C. E. and May, J. W. (1972). *Surf. Sci.*, **29**, 60, 85.

Ertl, G. and Koch, J. (1972). In *Adsorption–desorption phenomena* (ed. F. Ricca), p. 345. Academic Press, London.

Frenkel, J. (1946). *Kinetic theory of liquids*. Clarendon Press, Oxford.

Garmon, L. B. and Lawless, K. R. (1970). In *Structure et proprieties des surfaces des solides*, p. 61. Editions du Centre National de la Recherche Scientifique, Paris.

Gibbs, J. W. (1928). *The collected works of J. Willard Gibbs*. Longman, New York.

Gomer, R. (1967). In *Fundamentals of gas–surface interactions* (eds H. Saltsburg, J. N. Smith, and M. Rogers). Academic Press, New York.

Joyce, B. A. (1985). *Rep. Progr. Phys.*, **48**, 1637.

Joyner, R. W. and Somorjai, B. A. (1973). In *Surface and defect properties of solids* (eds M. W. Roberts and J. M. Thomas), vol. 2, p. 1. The Chemical Society, London.

Kelly, M. J. and Nicholas, R. J. (1985). *Rep. Progr. Phys.*, **48**, 1699.

Kittel, C. (1986). *Introduction to solid state physics*. Wiley, New York.

Lord, D. G. and Prutton, M. (1974). *Thin Solid Films*, **21**, 341.

Matthews, J. W. (1975). *Epitaxial growth. Parts A and B*. Academic Press, New York.

Mayer, H. (1971). In *Advances in epitaxy and endotaxy* (eds H. G. Schneider and V. Ruth), p. 63. VEB Deutsche Verlag für Grudstoffindustrie, Leipzig.

Newns, D. M (1969). *Phys. Rev.*, **178**, 1123.

Norskov, J. K. (1990). *Rep. Progr. Phys.*, **53**, 1253.

Pashley, D. W. (1970). In *Recent progress in surface science* (eds J. F. Danielli, A. C. Riddiford, and M. Rosenberg), vol. 3, p. 23. Academic Press, New York.

Pendry, J. B. (1974). *Low energy electron diffraction*. Academic Press, London.

Plummer, E. W. and Bell, A. E. (1972). *Proc. International Conference on Solid Surfaces*, p. 583. American Vacuum Society, New York.

Plummer, E. W. and Young, R. D. (1970). *Phys. Rev.*, **B1**, 2088.

Sandejas, J. S. and Hudson, J. B. (1967). In *Fundamentals of gas–surface interactions* (eds H. Saltsburg, J. N. Smith, and M. Rogers). Academic Press, New York.

Somorjai, G. A. (1972). *Principles of surface chemistry*. Prentice-Hall, New Jersey.

Vermaak, J. S. and Henning, J. A. O. (1970). *Phil. Mag.*, **22**, 269.

Watson, P. R. (1987). *J. Phys. Chem. Ref. Data.*, **16**(4), 953.

Weston, G. F. (1985). *Ultrahigh vacuum practice*. Butterworths, London.

Zangwill, A. (1988). *Physics at surfaces*. Cambridge University Press, Cambridge.

Index